Praise for The Missing Years.

(Finalist in the Australian Society of Travel Writers' Best Travel Book of the Year Award 2009)

'Absolutely compelling.' -- John Kerr, Radio 2UE.

'A very, *very* gripping story ...' -- Graham Mabury, Radio 6PR.

'An amazing read ... a historical document, which Stu Lloyd has done a fantastic job on.' -- Simon Owens, Radio 3AW.

'The Digger's memoir is fast becoming one of the most popular books of the military genre and several clamour for attention. Three of the best are *Dangerous Days*, *On Paths of Ash* and *The Missing Years ...*' -- The Weekend Australian.

'A very personal story that will get the hairs on the back of your neck standing up.' -- *Australian Men's Fitness*.

'A fascinating story ... a great book.' -- *Reveille* magazine.

'A rollicking good read. Gosh, what an incredible story. How anyone survived that catalogue of disasters and lived to tell the tale beats me.' -- Julie Summers, author *The Colonel of Tamarkan* and *A Stranger in the House*.

'A good and detailed account of the period. Pilkington had a good eye for detail. The power of the narrative is that it *IS* first hand. I read it pretty much in one go.' -- Lynette Ramsay Silver, author and military historian.

'A good and interesting read. We'd like to buy *two more copies* for the TBRC library.' -- Terry Manttan, Thai Burma Railway Centre.

'I'm hooked. It's an important work that probably needs to be a school text. The research and detail, linking the personal to the sweep of history, is just amazing and equally moving.' -- Peter Morrison.

'Your extremely interesting book puts an entirely different view of Changi and the railway than I had read previously. The views of victims of the Japanese are interesting.' -- Roy Finney, Military Vintage Rifle Club.

'We are in the hands of a truly professional wordsmith ... I am lost for words.' -- Rear Admiral (ret'd) Ian Richards, Royal Australian Navy.

Also by Stuart Lloyd

A Bleeding Slaughterhouse – The Outrageous True Story of The Alexandra Hospital Massacres, Singapore, 14-15 February 1942.

The Depths of December -- The Sinking of the HMS *Prince of Wales*, *Repulse*, and The British Empire, 1941.

The Grubby Little Men Who Raped Hong Kong -- The True Story of the St Stephen's College Massacre, December 1941.

The Malay Experiment – The True Story of How Captain Bruce Made the Malay Regiment Fit For Fighting, 1933-1942.

The Nightmare of Nanking – Selected Stories from The Nanking Massacres of 1937-38.

Tales from the Tiger's Den – An Oral History of Foreigners in the Far East from 1920 to 2020.

Visit amazon.com/author/StuLloyd

THE MISSING YEARS

A POW's STORY FROM CHANGI TO HELLFIRE PASS, 1942–45.

Stuart Lloyd

Until Death, all is Life.

— *Sancho Panza*

Never before have so many been fucked about by so few

And neither the few nor the many have fuck all idea what to do.

— *M.C. Hay, Volunteer, Malayan Civil Service*

CAT MAT DOG

CatMatDog Storytelling

First published in Australia in 2009 by Rosenberg Publishing Pty Ltd

This edition updated in December 2021 and published by:

CatMatDog Publishers, Australia.
ABN: 85312594003
catmatdog.com/stuart-lloyd-author

Contact the author: stulloyd.worldsmith@gmail.com

Cover design and photo pages: Stuart Lloyd

Interior typesetting: Shahid Aziz.

ISBN: 978-0-6453280-4-2
Available in paperback and eBook.

Contents

Introduction

Even Somerset Maugham at his gin-addled best would have struggled to conjure up a story more compelling, more emotionally evocative, and ultimately more disastrous than the battle for Malaya and the subsequent horrors endured by the prisoners of the Japanese.

As a universal backdrop, the overarching theme of Empire versus Emperor, riddled with racial motive and overtone, set against the steamy tropics of South-East Asia. Add a supporting cast of thousands, *tens* of thousands, of Western allies, allied in cause but often with national differences — let alone class differences between officers and other ranks — sniping away in the wings. And front and centre, Oriental villains so dastardly as to be almost implausible, pitted against a dashing strong-jawed British army captain, Hugh Patrick Pilkington.

Pilkington typified the British Raj in the Far East in many ways. His family had long since sold out of their interests in the Pilkington Glass business, his father working as an engineer with Indian Railways when Hugh was born in Madras in 1904. The Raj was in full swing, at its zenith of imperial influence. Young Hugh was sent to Shrewsbury (one of the top public schools in England), where he captained his school's first rowing eight at Henley, and was generally just the right sort of chap to be plucked by Dunlop's for a plum position as a rubber planter in British Malaya in 1922. There he lived the stereotypical 'whisky-swilling' planter's life (a phrase coined by *The Times*' correspondent Ian Morrison) for fifteen years, topping out at seventeen stone on a daily diet of curries, supplemented by doughnuts according to the 1927 Malaya Cup rugby program.

In 1932 he married Phyllis in Kuala Lumpur. Looking for a change in his life more than hers, they arrived back in England in 1937, and — as war consumed Europe then rapidly more of the globe — he soon found himself in the Territorial Army. Of all the possible theatres to which his unit could be assigned, he amazingly found himself dug-in within 100 kilometres of where he had lived in Malaya.

A Japanese sniper cut him down on the battlefield, consigning him to a military hospital in Singapore, which was subsequently overrun by the rampaging Imperial Japanese Army. This culminated in a heinous massacre just hours before the capitulation of Singapore.

This episode turned into a morbid fascination – or obsession, some may say – as I dived ever deeper down its rabbit-hole for the next decade after I'd written this book. The result was unearthing tons of new research and mind-blowing unpublished memoirs and manuscripts from survivors, which fed into my book *A Bleeding Slaughterhouse*, published late 2021. That new research has found its way into this edition, and updates -- and, yes, corrections -- where required have been made and incorporated here.

Subsequently then for Pilk came the rigours of life as a prisoner of war for three and a half years, including being chain-ganged on the Thai–Burma 'Death Railway' with H Force, before liberation eventually, finally, thankfully, came.

But with his battle injury and overall physical debilitation, plus the inevitable mental scarring, was he ever truly liberated?

This then is the story of Captain Hugh Patrick Pilkington, 96905, of the 6th Royal Norfolks. More than simply a documentation of one man's living hell, this is an exploration of the ups and downs of Empires, politics, society, family, mateship. Part military history, part travelogue, pure human spirit. Viewed originally from the safety of 60 years' (and now 80 years') distance since the Fall of Singapore.

In fact, this story very nearly never saw light of day. You see, it was at pain of execution that Pilk (as he was familiarly known to friends) kept his secret records on scraps of paper often purloined from the Japanese themselves, adding an extra illicit frisson. It became a memoir when compiled on board the POW repatriation ship HMT *Sobieski* as she steamed from Singapore on 27 September 1945 back home to the UK. The details were vividly recalled, and unfiltered by the dimness and distortions of time. Raw and accurate. Unaffected by what became known as 'Changi Memory

Syndrome' in which details were lost and episodes blurred unreliably into one another.

The Sancho Panza quote, repeated here, was emblazoned on the front page. Dedicated to his wife Phyllis, Pilk's war memoirs opened by saying: 'In order to give you some idea of the missing four years of our lives, I am writing you this short account'. This allows us, the reader, a voyeuristic fly-on-the-wall peek into the most private moments of the man's mind. His son, Paul Pilkington, told me the bound manuscript sat on their familial bookshelf in England for years. 'But no-one ever read it.' Paul also told me how these memoirs had subsequently gathered dust in his attic in Sydney for years, along with various other inheritances from his dad: a Malay Regiment forage cap, some ceremonial *krisses.*

It was only one hellish summer in bushfire-besieged Sydney that Paul — living adjacent to national parkland — gathered up everything in preparation to evacuate his home. There was *that* box containing the folder entitled 'THE MISSING YEARS 1941–1945, being the diaries and letters of H.P. Pilkington'.

In 'The Missing Years', Pilk made frequent reference to the fact that it should be read in conjunction with his letters home from various training camps in England and Wales, and along the oceans of the world, for the full story. Paul was initially reluctant to share these, saying it was 'private family stuff '. However, my increasingly desperate entreaties finally paid off, and Paul delivered another folder of now-typed letters and addenda.

Further interrogation led to him delivering boxes of photo albums, newspaper clippings, telegrams, postcards. The resulting assemblage makes one realise how damagingly disjointed war can be. No news. Out of date news. Wrong news, rumour, hearsay. False dawns.

There were also manuscript drafts, and several leather-bound volumes with gold-embossed lettering. Inside, proudly printed, was 'Ex-libris H.P. Pilkington'. I flicked through one, scarcely believing my eyes: page after page of Ronald Searle cartoons, caricatures and watercolours. (Searle, the renowned English cartoonist, was a fellow H Force and Sime Road camp-mate of Pilkington's.) Another manuscript, *Bungalow Days*, recounted Pilk's life as a rubber planter in Malaya, where servants, tigers, dogs, pigs, snakes, and cows came to life in the Malayan idyll under Searle's steady hand.

Additionally, there was a well-worn cash book detailing virtually every dollar earned and spent by Pilkington in Malaya between 1922 and 1937, along with Planters' Club chits from all parts of western Malaya, and photographs of the high-living English bachelors taken invariably by a Japanese photographer, Kunita, in Seremban, capital of the central Malayan state of Negri Sembilan.

How valuable was Pilk's experience of the tropics, the terrain, even the local languages, to the captain and his men — many of whom had probably never even seen an Asian back home — when he found himself fighting for his life in the very area where he'd worked for so long? And what bittersweet feelings would have been evoked as he saw again the adopted country he loved under such vastly different circumstances?

It also made me wonder, among growing nostalgia and 'celebration' of war veterans in western countries — not least in Australia — what the locals in Singapore, Malaysia and Thailand make of this war brought to their lands by the Japanese desire to rid them (especially in the case of Singapore and Malaya) of the white colonial yoke and establish the Greater East Asia Co-Prosperity Sphere: Asia for Asians. And what a well-intentioned though horribly misguided veneer that proved to be. The Japanese simply wanted to be the new 'whites' of an Asia they had been beaten to by the European colonial powers. The local Asian populations have all but been overlooked in the telling of this story by Western and Japanese writers. Importantly, what are their views towards the Japanese today?

My powers of empathy are unusually strong but I failed miserably to empathise with Captain Pilkington, his men, and fellow POWs. Episode after episode of unfathomable adventure and atrocity: any one of the episodes would have been enough to finish me off. I couldn't possibly survive on the 'Railway Rations' of rice just typing at my computer, let alone breaking rocks in the blistering tropical sun at bayonet point for twelve hours or more each day, for months on end. *Could I?*

That set in train another dimension of this undertaking: to retrace with Paul his father's steps through Singapore, Malaysia and

Thailand, a life-changing experience for both of us, detailed in the epilogue of this book. I think in the 14 years since, Paul has almost forgiven me. *Almost!* Another fulfilling sidebar to this is that I've managed to introduce Paul to relatives of others contained in his father's story, such as Angus Robertson, the grandson of his former commanding officer, Lt Col Ian Lywood. Everybody gains from the casual collisions of these circles.

I have added contextual clarification to Pilkington's manuscript to aid and abet the understanding of local events in the wider context of the war that he may or may not have known about, and the occasional elaboration of wisdom with hindsight. One facet repeatedly jumped out at me — the better the war was going for the Allies, the worse it got for the POWs.

By war's end, despite an Allied victory, the unthinkable had become inevitable: the sun had set on the Raj. And no one, least of all Somerset Maugham, could have seen *that* writing on the wall in the pre-war idyll. But thankfully Pilk lived to tell his side of the sad and sorry tale. Lest we forget.

Stuart Lloyd
Orange, NSW. December 2021

Captain Hugh 'Pilk' Pilkington, 6th Royal Norfolks,
aged 35 in 1939.

Prologue

A WORLD OF FEVER AND PAIN

From the memoir of Captain Pilkington, Alexandra Military Hospital, Singapore, February 1942:

'From my bed I had a splendid view of the battle of Singapore. I could watch the docks and town being bombed, the fighting on the hills, on the island. In fact there was little I missed.

The officers' wards were at the top of the hospital on the third floor, open on both sides with broad balconies and very cool all day: a continuous draught blew over me.

My temperature was seldom below 103 and generally well over 104. I lived in a world of fever and pain with happy hours in between when I had Morphia and sleeping draughts. My wounded shoulder was suppurating so badly that not only was it padded with cotton wool six inches deep but thick bath towels were placed underneath me and these had to be changed every four hours. I was gradually able to use my free arm and legs and do little things like feeding myself and lighting my own pipe.

On Wednesday 11 February some 500 patients from another hospital that had been bombed were moved in while the hospital was under fire. We were now so thickly packed that patients were lying on mattresses in the passages and on the balconies.

Twenty-five more sisters were evacuated (yesterday, fifty of the sisters, much against their wills, were evacuated), and the hospital had been completely deserted by its Chinese and Eurasian staff who had bolted to a man.

The hospital was continuously under shell and mortar fire with the doctors and sisters working in steel helmets. Tiles and roofing were flying all over the place. It was terrible to hear the howls and shrieks of the shell-shocked patients every time a gun went off.

A direct hit from a mortar bomb on the tiled roof above our ward caused one of the sisters to raise her helmet and say, "Come in. You needn't knock so loudly, I'm not deaf!"

A new bed had been prepared for me. A wooden fracture bed with an ordinary mattress, on top of which was a big Dunlopillo one with linen sheets and huge feather pillows. A big argument took place as to the best method of getting me into bed, aggravated by my remark that I weighed 16½ stone without the masonry. However, four orderlies and three sisters rallied round and I was lifted most gently into bed. I have never known anything so comfortable, and veritably floated on air. Underneath my bed, with its two mattresses and my lovely plastered body, was a refuge for all during the air raids and it always surprised me the number of people who could get under it in an emergency!

I was operated on again during 12 February when my plaster cast was renewed in a desperate attempt to get me away on a hospital ship, but it was not to be. They considered me far too weak and said I could never survive the handling and journey.

We were on a liquid ration of half a coffee cupful of water three times a day as the water main had been hit. Fortunately I had a large thermos flask full of water as well as some bottles of soda which kept my feverish thirst at bay for a bit. Our food was now cut down to two meals a day of one piece of stale bread and marmalade.

By midday all patients who were fit to walk had been told to find their way to Singapore.

Later on, a heavy raid was made on the oil tanks in the vicinity of the hospital and, although we had been told that they were all empty, they succeeded in burning for three weeks, making the night ghastly with their red glow while, during the daytime, great clouds of black smoke drifted through the hospital, making it sometimes difficult to breathe.

Friday morning showed that the electric power lines had been hit so that now, added to our thirst, was the intense heat of a room full of feverish patients, with no fans working and no doors open. The rest of the sisters were evacuated.

Friday night passed in an inferno of noise and light, but now our thirst was nearly unbearable with many officers delirious.

On the morning of 14 February, the battle was raging in the

hospital grounds, with our own artillery now searching for the Japs amongst the oil wells [Normanton oil tanks]. At 14:00 the Japs put in a full-scale attack in this area driving back the British and Indian troops, who withdrew through the hospital wards. For an hour they were firing from the balconies and windows but were gradually driven out. A great many patients had been killed. The Japanese had thrown hand grenades wherever there was resistance, smashing theatres, dispensary and kitchen, and killing patients, doctors, and even bayoneting people on the operating tables.

By 15:00 the hospital was in Jap hands and — except for an occasional bullet from one of our snipers — the battle had died down.

The Japs now took matters into their own hands. They paraded every walking wounded, marched them out of the hospital, tied them in sixes to the trees near the burning oil wells, bayoneted them and threw their bodies into the fire. One officer escaped to tell the tale.

Next they removed all the doctors and orderlies and divided them into two parties. One lot of 100 they locked up in a barrack block near the hospital without water and food, collecting them the next day in batches of ten, when they were taken to the rear and mown down by machine-gun fire. Two orderlies escaped to tell the tale.

The two hundred remaining doctors, orderlies and hospital staff were marched off to a deep drain at the rear of the hospital where their hands were tied behind their backs and they were kept till 12:00 next morning without water and food, when they were allowed to return to the hospital till 17:00 and the procedure was again repeated on Sunday night.

By now the hospital was denuded of everyone except those absolutely bed-ridden. We were warned that there must be no smoking or talking.

The night was a ghastly nightmare. The flashes of the guns turned the night into almost blinding daylight which, with the burning oil wells, would have given Dante a good idea of a real inferno.

People tried to quieten delirious patients. No one in our ward could do more than move about slightly in bed and there was no one to carry out the normal sanitary procedure with the result that, to the stench of suppurating wounds, was added every type of smell.

I was trying to read *Gone with the Wind* and naturally had got to the part where the hospital was under fire.

By nightfall all the stray dogs of Singapore appeared to be in the hospital and rushed, barking and growling, up and down the wards. One officer had had an abdominal operation for which a tube was used to drain the wound. The dogs got hold of one end of this and a tug of war then ensued. The dogs eventually won.

For 24 hours we had no food or water. At 12:00 next day the orderlies and doctors still surviving were allowed to return. In a few hours they worked wonders and by 17:00 we had all had a meal of sorts, were cleaned up and comfortable.

During Sunday 15 February our own guns were ordered to fire on the hospital and they proved to be far more accurate than the Japs — with a few salvoes they completely demolished the top floor of one wing and removed most of the roof elsewhere. Two shells burst on the stairwell by our ward and for two hours bits of shrapnel and ceiling rained down on us like snow. Added to the inferno were the choking thick fumes of cordite.

At 18:00 all firing stopped and we heard the sirens sounding the All Clear, immediately followed by the "*Banzais*" of the Jap soldiers. An eerie calm then fell across the island. Singapore had fallen.'

Curiously this episode, which became known as Alexandra Hospital Massacre, does not rate a single word in *History of the Second World War: The War Against Japan* by S. Woodburn Kirby, the official British history. Its Australian counterpart, *Australia in the War of 1939–1945: The Japanese Thrust*, by Lionel Wigmore, affords the episode a scant half dozen lines.

But unbelievably, worse — *much worse!* — lay ahead for Pilkington and his beleaguered men who were now prisoners of war of the Imperial Japanese Army.

1 MARCHING ORDERS

England and Scotland, 1937–41

So how exactly did Captain Hugh Pilkington come to be lying in a hospital bed in Singapore? Let's backtrack a little ...

Pilkington had undergone ten years of diligent training with the Malay States Volunteer Rifles in Malaya since 1927. Malaya was then effectively controlled by 200 British Police but volunteers were needed to make up for the perceived military deficiency. (Only one battalion in peninsular Malaya existed at the time: the Burma Rifles at Taiping, with one battalion, the Wiltshire Regiment, stationed in Singapore.)

The Malay States Volunteer Rifles was the first to set up, in 1915. Later came the Federated Malay States (FMS) Volunteer Forces and the Straits Settlement Volunteer Forces. Brits were expected to join. 'It wasn't at all compulsory but the system and social set up was such that you would not have been comfortable if you hadn't,' said Edward Tokeley in *Tales from South China Seas*. As the name suggests, they were not paid but rather reimbursed for petrol. Pilk often claimed over $20 (Malay dollars) per month in the later pre-war years, a fair sum given reimbursement at 20 cents per mile.

On 1 March 1933 an experimental unit was formed with the aim of seeing how the usually easy-going Malays would react to 'military discipline'. The 25 recruits acquitted themselves well under Commanding Officer Major George Bruce of the Lincolnshire Regiment.

The 150-strong Malay Regiment (*Askar Melaya*) officially came into being 1 January 1935 in Port Dickson on the coast near Seremban where Pilk enjoyed many weekends of sailing. Their motto was *Ta'at Setia*, which can mean 'Loyal and True' or, more colloquially, 'Last Man Standing'. Their logo was *harimau berjuang*, the fighting tiger.

Pilk attended the often-casual weekly meetings, sometimes a weekend camp. It was all treated as a bit of a lark: they were the unassailable mighty British who ruled the waves and held sway in much of Asia for 80 years, after all. Photographs show him at machine-gun practice in pith helmet, while officers stand around in Sam Browne's, ludicrously roomy shorts (nicknamed 'Bombay Bloomers' or 'Empire Builders'), and long white socks, peering through oversized binoculars. The weapon was a World War I-model Vickers. Locals in the Malay Regiment were soon promoted to platoon commanders, training with the Royal Inniskilling Fusiliers in Singapore in 1936.

Those with prescience seemed drowned out by the overall complacency. 'I remember so vividly one exercise which was co-ordinated throughout the country by the GOC, Major-General Dobbie,' recalled Tokeley. 'When it was over he gave us a lecture — and I make no mistake in what he said then. He said, "the enemy will be Japan and the Japanese will come down the mainland".'

Thirty-six-year-old Pilk returned to England after fifteen years enjoying the 'happy, carefree existence' of a rubber planter in the central state of Negri Sembilan, Malaya. Back, he wasted no time getting stuck into his favourite sports — hockey, tennis and squash — and Phyllis, whom he'd married in Kuala Lumpur in 1932, busied herself in keeping house and dabbling with landscape painting. They chose Norfolk in East Anglia as ideal, with its sparse population, abundant waterways and birdlife.

Hitler soon pounced on Austria, and China was also at war — Tsing-tao, Canton and Hankow had now been occupied by the Japanese. After the Jinan and Mukden incidents, flashpoint for Japan's all-out invasion of China, was the Marco Polo Bridge incident of 7 July 1937. (The Lugou Bridge, as it's also known, guarded the only passage linking Beijing to National Revolutionary Army-controlled areas from the south. If the bridge fell, Beijing would be completely cut off and easily captured by the Japanese.) This marked the official commencement of the Second Sino-Japanese War, as part of the Japanese plan to control the Asia mainland.

Japan looked to Britain as a role model: both were small island nations. But Britain boasted an enviably large outpost Empire. It was unanimously accepted that a strong navy was a minimum prerequisite.

In England in 1938, the order was given to double the size of the Territorial Army (the volunteer reserve force of part-time soldiers). The Royal Norfolks added two battalions, 6 and 7, duplicates of the existing 4 and 5 Battalions.

What news the Pilkingtons heard from Malaya was fairly upbeat, given that rubber was cruising on a nice upward trend, above the historically lowest prices of 1932 again for the first time. However, there was unrest among the plantation workers clamouring for a better share of the lucrative output. The nationalistic Chinese campaigns, a pro-Chinese move in retaliation for Japan's expansionist overtures in China, seemed to be holding, with Japanese exports to Malaya dropping by 75 percent from 1937 to 1939.

In that latter year, sports-mad Pilk — who had represented his adopted Malayan state at rugby, hockey and soccer — was appointed as Secretary of the Norfolk and Suffolk National Fitness Council. On 1 September 1939 the 6 Royal Norfolks mobilised as a battalion in Norwich. Pilk signed on as an HO (Hostilities Only) non-regular. The day German troops reached Warsaw, he was gazetted as Lieutenant Hugh Pilkington, 96905.

Over in Singapore, the Straits Times broke the news that Britain was at war and opined: 'At this distance from the scene of battle, with our defences perfected and Japanese participation in the struggle on the side of Germany an extremely remote possibility, Malaya has little to fear.'

In October 1939 Britain conscripted 250, 000 more troops. (Australia, by contrast, was for much of the war the only major combatant nation that had no conscription yet a full 50 percent of males aged 18 to 45 signed up.) Numbering just over 500 in all ranks, the 6 Royal Norfolks were divided into three rifle companies, but when the swelling numbers reached 785 a 'D' Company was formed on November 30. Year-end saw them shift to Sheringham. The 6 Royal Norfolks were brigaded with the 5 Royal Norfolks and 2 Cambridgeshires to form 53 Infantry Brigade, in turn part of 18 Division.

The winter of 1940 was particularly severe, and the brigade commander Lieutenant Colonel Buxton offered his men to the Norfolk County Council to clear the massive snowdrifts. While this was hard work, their field training and route marches suffered. Confined largely to halls such as the dining room of the Sheringham Hotel, daily lecture sessions were run from 09:00 to 15:40, with one-and-a-half hours for lunch from 12:30. Pilk conducted classes including war gasses, fire control and orders, section in defence, camouflage and range cards, and 'Gas — decontamination of personnel and weapons. Practicable.' Further training in the handling of rifles, anti-tank rifles and Bren guns was also given — turns being taken to handle the mere six Bren guns and single two-inch mortar the battalion could muster from its armoury.

Over in Malaya on 1 May 1940 the British Police decided they'd had enough and raided the Malayan Communist Party leaders: the strikes and disruptions to trade they were masterminding had caused rampant inflation. One thousand were rounded up and deported to China. Britain could ill afford such shenanigans — after all, the Governor of Singapore estimated at the time that Malaya was worth an estimated £227.5 million to the Empire. Exports alone were valued at £131.25m, £93 million of which went to 'foreign' countries other than Britain — chiefly USA and Canada. It was a vital dollar arsenal to keep Britain's war effort supplied with rubber and tin (of which Malaya supplied around 40 percent and 60 percent, respectively, of the world's total output). By July, the Communist Party had done a backflip, ceasing its struggles against British colonialism.

October 1940 saw one Robert Brooke-Popham appointed as British Commander in Chief in the Far East. The Sandhurst graduate and Boer War veteran was immediately previously governor of Kenya, a position to which he had retired, having not found favour as Air Officer Commanding, Fighting Area, where he was in charge of the UK's entire air defences. While he enjoyed a pompous-sounding title, his Far East appointment was really to adjudicate quarrels between the Army and Air Force. Furthermore, he had absolutely no authority over the Navy.

The following month, with Brooke-Popham desperately trying to build up the defences of the South-East Asian region, the British ship SS *Automedon* was sunk en route to Singapore by a German raider. On board was a pessimistic document on defence appreciations by the Imperial General Staff which indicated that

Britain was unable to reinforce Malaya. These highly classified documents were intercepted by the Germans and passed on to the Japanese.

The reality of the situation was that the Army Commander for Malaya, Lieutenant General Lionel Bond, estimated that between 39 and 42 battalions were required to defend Malaya, which was approximately the same size as England (although with a considerably smaller population of just five million). Lieutenant General Arthur Percival had just ten infantry brigades at his disposal, and not a single tank. Further, Bond estimated 336 first-line aircraft would be required. As it stood, as of September 1940, only 90 first-line aircraft were available — and those were mainly aging Brewster Buffaloes, unwanted by the Air Force for their war in Europe.

Back in England the whole of 18 Division was brought into reserve at Swaffham. Every available house and farm was occupied by these now-restless soldiers who saw out the year there.

January 1941 bought with it a move to Scotland for the Division for further advanced training. Scotland was chosen because its barren treeless plains best replicated the featureless terrain they might encounter in the Middle East, specifically Libya. There was little doubt in anyone's mind that was where they were bound. After all, Churchill had just announced that, 'The political situation in the Far East ... does not seem to require the maintenance of large forces in the Far East at this time'. The Scottish weather proved a formidable foe, with snow almost daily in January and February. Smaller company-scale exercises could proceed, but the larger brigade- and divisional-level exercises had to wait until March.

Japan went public with its expansionist ambitions on 2 July 1941. At the Imperial Conference in Tokyo they agreed to the construction of the Greater East Asia Co-Prosperity Sphere. French Indo-China, Malaya, the Philippines and the Dutch East Indies all possessed the much-coveted natural resources required to fuel the fledgling Empire's expansion. Notably, they were all under Western colonial masters. Trade within the 'Yen Block' (essentially Japan, Korea, Manchukou and parts of South-East Asia) was considered successful, representing between 75 and 80 percent of their trade at that time. But the war on China severely warped Japan's trade balances to the point of self- strangulation. Japanese forecasts at this time show that they would have exhausted their currency reserves by spring 1942. America, which supplied 60 percent of Japan's oil, imposed trade embargoes on Japan in July 1941.

July 15 saw command of the Norfolks change hands once more, with Lieutenant Colonel I.C.G. Lywood taking over the reins from Lieutenant Colonel F.L. Cubitt. August had the division on the move again, this time to Knowsley, Lancaster, a few kilometres from Liverpool. The battalion was put up at Knowsley Park. While luxurious lodgings existed in the form of Ormskirk and Huyton Lodges, the Norfolks were not so fortunate, sleeping under canvas until more permanent Nissen hut-type digs could be erected. 'Our camp at Prescott, Lancashire, was in Lord Derby's beautiful park. Deer, sheep and cattle roamed at will. Right up to the end I was in a tent near a huge oak beloved by the deer.'

The Norfolks' fitness dwindled: Little or no training could be carried out in this congested manufacturing district, and long route marches of twenty or thirty miles a day that had done so much to make the men super-fit, were impossible over the cobbled streets, choked with heavy traffic, of this district.

But it was deemed that 6 Royal Norfolk Battalion, 53 Brigade, and 18 Division were ready for action abroad.

Phyllis, with her mother and a Miss King to look after her, was installed in Woolacombe, North Devon: 'A lovely house called "The Castle" which is surrounded by lovely big walls and cliffs and things.' It was kindly lent to them by Phyllis' aunt, Ethel. Silver, the family's new pet dog, accompanied them on walks through the Victorian resort area which had provided the setting for books by both Agatha Christie and Arthur Conan Doyle.

On October 22, 1941 King George V1 conducted an inspection of 53 Brigade at Crewe Hall, Crewe.

'The King spoke to nearly every man. I have never been so impressed by personality before. His frank blue eyes and gentle smile and his few kind words to us made us feel not only that he was really a king, but *our* king.'

Four days later, in a rain-sodden canvas tent in Prescott, Liverpool, Pilk penned the following letter home to his six-and-a-half-months- pregnant wife, Phyllis:

'It rather looks as if we shall be off any day now. Will you please look after the enclosed receipts, driving licenses, etc., for me and wind my watch up at least once a week. Old Moore says we shall be demobilised next September and so does Lindoe, so I hope they are right. Don't worry about me. I shall be perfectly all right getting fat on bananas and curries and mangoes but do look after yourselves and Silver as I am longing to be home with you all again.

My future address will be: 6 Bn. The Royal Norfolk Regt. c/o Army Post Office No 1635.'

Pilk's optimistic naiveté seems designed to placate Phyllis as much as buoy his own spirits that this trifling adventure would soon be dealt with, and the routine of family life — with the exciting addition of a child — would resume again shortly. It is doubly poignant and ironic against the bleak and austere conditions of an English wartime autumn, and the shocking reality of what was shortly to unfold.

With each day Pilk's anticipation grew, their destination still unknown to them. By now a captain, he was writing letters home daily:

27.10.41

'Don't forget to write at least three times a week and number your letters so that I shall know which one to read first!'

28.10.41

'C and D Companies were the last to leave our camp at Prescott. Fortunately, while waiting in the mess we found 1/2 gallon of rum the quartermaster had left, and with this and what was left of our mess stocks, we passed away the hours. The noise made in the middle of the night by 300 men moving through the inky black-out laden with everything, including their weapons, kitbags and tropical helmets, was enough to waken the dead. Dropped helmets proved excellent footballs and generally, as a secret move, it was a failure. The only thing I am glad about is to get out of that filthy camp. We left it covered with thick glue-like mud, pelting with rain and all our tents soaking wet. I think another few weeks of it would have been unbearable.'

The 6 Royal Norfolks embarked at Gourock, near Glasgow, on

28 October 1941. The HMT *Duchess of Atholl* was a double-funnelled, trans-Atlantic Canadian Pacific Line ship of 25, 000 tons.

'We sailed at 11:30pm on October 30, my birthday, and the following morning, escorted by six destroyers and an Ack-Ack [anti-aircraft] cruiser, we met up with the rest of our convoy containing the balance of 18 Division. In all we were nine ships.

On the morning of 4 November we woke up to find the sky full of American aeroplanes while on the horizon was a convoy of about 50 ships. Later in the day the American fleet, which consisted of two battleships, four cruisers, the aircraft carrier Lexington and twelve destroyers, took over our convoy and we said goodbye to our little destroyers. In spite of the huge size of the American fleet I am sure we felt much safer with our six destroyers.

It should be remembered that this was the first convoy in history of fighting troops to be escorted and taken under the protection of a neutral power, ostensibly at peace with Germany.'

5.11.41

'We all settled in on this luxury liner very comfortably. The food is really marvellous; not a great deal of choice but at least four or five courses to choose from. The boat seems to have been completely stocked with colonial produce; South African tobacco and wines, Canadian chocolate and sweets, etc. All very plentiful and cheap. Everyone seems to be trying to smoke and eat themselves silly.

I share a table with Monty Smyth, Cubitt, Campbell-Orde and the Padre. They were much impressed to see that the steward always brings me a double ice-cream, but haven't the courage to make a similar arrangement.

The day before reaching Halifax [in Canada] we were joined by another twenty-four destroyers of the Royal Canadian Navy, so that by now, as far as one could see, were naval craft.'

8.11.41

'Last night we had strawberry ice-cream with real strawberries and for lunch a pre-war mixed grill! I wish I could send Silver some of the millions of bones and rindy skins we have over, but I fear it would not be too popular with the P.O. The seas round Halifax were particularly dangerous, four merchantmen being sunk just outside the harbour a few days previously. There were one or two scares when depth charges were dropped, but the only kill was a whale and a school of fish.'

Unknown, and seemingly irrelevant to Pilk and his men, a game of musical military chairs was in progress in Japan. On 6 November 1941 Yamashita Tomoyuki was recalled as commander of 4 Infantry Division in China. He had fallen foul of the Showa Emperor, and also Tojo, by reasoning that Japan should maintain cordial relations with both England and America, and end the assault on China. He was relegated to command 25 Army — three divisions totalling 60, 000 men in reserve on Hainan Island, China — and was now literally fighting for his honour and his career.

'Halifax, reached about November 9, was a great surprise to us as it was all lit up and made the war feel far away. The ship gave us a magnificent farewell dinner of all the luxuries we hadn't seen since the outbreak of war.

After we had berthed a convoy of about four hundred ships, which had assembled in the harbour, started to steam out. They went by about one every two minutes and took about twelve hours to clear. This, we were later told, was the biggest convoy that had crossed the Atlantic. Quite a number were unfortunately sunk, some of them only a few miles outside Halifax.'

At Halifax they transferred from the *Duchess of Atholl* to the USAT *Mount Vernon*, an American trooper under Admiral Donald Beary. This ship had formerly been the SS *Washington*, the graceful top-line American trans-Atlantic liner. She was 29, 000 tons, luxuriously fitted and capable of some 24 knots.

The whole of 53 Infantry Brigade under Brigadier Duke — consisting of 5 and 6 Battalions Royal Norfolk Regiment, 2 Cambridgeshires, Brigade Headquarters, 135 Field Regiment Royal Artillery, 198 Field Ambulance, RASC Company, Provost Company 'and last but not least the bath unit' — boarded the Mount Vernon. In all some 5500 officers and men were squeezed into accommodation intended for 2000. The transfer from the British ship to the American was completed in 'the record time of twenty-four hours, with the result that next day we were steaming out of Halifax'.

The convoy consisted of three liners — the *Mount Vernon, West Point* and *Wakefield* — with three other older troopers of about

10, 000 tons. Unlike a British trooper, there were no mess decks, the men being accommodated in cabins where a two-berth cabin would be made to hold six.

'The Americans packed a British Division of 25, 000 odd men that, by British standards, required nine large ships, into about half that space. Deck space was too crowded except for 20 minutes PT. We continue to get a large and varied assortment of meals and — from all appearances, except for the blackout — one might be travelling out East again.

I shared a palatial cabin with Peter Hill and John Stratford, which was still furnished as in peace-time with down pillows and vi-sprung mattresses.

Feeding was on the cafeteria principle: the men were given a tray with little partitions and, as they passed down the serving tables, portions of meat, vegetables, soup, bread, and butter were all ladled into their correct place, with the result that they finished up with a complete meal. A curious point about this system was that a man wanting a second helping of, say, potatoes had to rejoin the serving queue and could finish up with another complete meal, most of which he didn't want!'

14.11.41

'The men are being really well fed and thoroughly appreciating it and getting heaps of eggs, butter, fresh fruit and even asparagus. All day long the men line up for their meals. As breakfast finishes, lunch begins!'

'The Americans, for the first ten days, carried out 'darkening ship,' which consisted of turning off all the light power except for a few blue lamps, so that even in the bowels of the ship, where there were no portholes, the men had to sit in almost complete darkness. 'All portholes were closed with steel dead lights for the duration, with the result that even in cold weather the men were so hot that they wore nothing but shorts.' This precaution was later relaxed and they adopted the British method of keeping all dead lights closed at night with special light trap exits to the decks.'

Until they reached Cape Horn, the convoy was escorted by two cruisers, a battleship, eight destroyers and an aircraft carrier, together with an oiler, all belonging to the American Navy. The first port of call was Trinidad, where no shore leave was granted: 'Trinidad has a lovely half-moon harbour, with sandy shores, deep tropical foliage reaching down to the beach edge with a glorious panorama of blue mountains as a background. It reminded me of Port Dickson and Penang fused into each other.' The latter two ports were in Malaya, and often frequented by the Pilkingtons.

20.11.41

'The food remains good and occasionally they regale us with a large double ice-cream to keep us quiet. It is very amusing to come across old friends on the menus such as papayas, pomelos, avocado pears (which I detest), melons, brinjals, and those string beans one got so accursedly tired of. Todhunter has a passion for avocado pears so I swap for his papaya. Bananas are very common!'

They followed the South American coast down to the approximate latitude of Cape Horn, where Pilk found it bitterly cold and not a little boring. 'Life has become a monotonous round of sleeping, eating and rushing for a bath. The food line is still quite interesting at times. We had pistachio nuts for dinner last night.' The pistachios also reminded him of Malayan days, where they grew on trees in his garden at Ainsdale Estate, Seremban.

'The boat appears to have run out of egg cups and we are now presented with our eggs, lightly boiled, rolling about on a plate. None of us had taken kindly to the US Navy's dry ship policy, with the result that the brandy from the Medical Comforts Hampers had by now run out. The men miss their beer more than anything.'

The *Straits Times*, Monday 1 December: **'ALL COLONY VOLUNTEERS CALLED OUT'.** Governor Shenton Thomas proclaimed a state of emergency, but the newspaper remained placatory on his behalf: 'It is emphasised in official quarters that ... a state of emergency does not indicate that there has been any further serious deterioration of the Far Eastern situation'. Meanwhile, Pilk's convoy followed the Antarctic Circle:

4.12.41

'Not much news except about food. We have had pumpkin pie, lime jelly and cranberry pie (called huckleberry pie) and now we are allowed two ice-creams each when we have them, which isn't quite so often. We couldn't get any chocolate at our last port of call which was a nuisance as we have eaten the boat dry, even of boiled sweets and biscuits.'

The *Straits Times*, Monday 8 December: **'JAPAN STRIKES AGAINST BRITAIN AND US'**. Japanese bombers staged an air raid of Singapore in the early hours of 8 December simultaneous with the attacks on Pearl Harbor, south Thailand and Kota Bahru, Malaya. The city was not blacked out.

Amazingly, Pilk's letter home dated 9.12.41 (written while at sea near the polar region) makes no mention of the 7 December Japanese attack on Pearl Harbor which was to profoundly change their destiny, bringing America into the war, and South-East Asia into sharp focus. Indeed, the Japanese assault on Malaya commenced with landings at Kota Bahru in the north-east, one hour and ten minutes before the attack on Pearl Harbor. Colonel Philip Toosey was with Brigadier Duke in the Captain's cabin of the *Mount Vernon* when the news of Pearl Harbor was broadcast. Perhaps that news was kept privy to the upper echelons?

'Nothing much to report since I wrote last. We have been having a lot of maize, sweet corn, and beans lately. They make a change but I definitely do not like sweet potatoes and yams.'

They reached Cape Town where a much-anticipated four-days shore leave was enjoyed to the fullest.

14.12.41

'We bought oranges 12 for 1, mangoes 2d each, and peaches 16 for 1 and some Cadbury's chocolate at 2 per pound. We then went to the Café Royal and had a couple of bottles of Wiesenbach white wine which we like very much and felt very mellow after it. On Table Mountain we had tea in a granite built chalet looking straight out over the mountain edge. We had real cream cakes, tea and this with fresh raspberries 1/6d. We went to the Del Monico Café and had drinks in the courtyard. We had hors d'ouevres, a *huge* mixed grill and a Del Monico Special for 7 (including beer). I hadn't had a gorge like that for years. Cook and I then set off for the cinema. After the cinema we made our way home, buying a supply of peaches to keep us going through the night.'

15.12.41

'Today we heard of the sinking of the *Duke of York* [*sic*] and

Repulse at Singapore.'

The two capital ships of Vice Admiral Tom Philips' Z Force, the *Prince of Wales* and the *Repulse*, were sunk on 10 December in an unprecedented incident by land-based Japanese aircraft, off the east coast of Malaya. The *Wales* was one of Navy's most modern battleships and the *Repulse*, launched in 1916, one of its oldest battlecruisers. Four destroyers — but no aircraft carrier, which had been damaged in the Caribbean — made up the numbers.

Churchill's reaction summed up what those in Britain and South- East Asia felt about this devastating blow: 'In all the war ... I never received a more direct shock. As I turned over and twisted in bed the full horror of the news sank in upon me.' May well he have been shocked — it was on his personal instruction, against the will of the Admiralty, to divert Z Force to Singapore. A show-of-force stunt gone horribly awry.

'The ship was full of rumours of a change in our destination but, as we were not consulted, we continued to enjoy ourselves.'

With Phyllis now eight months pregnant, Pilk's obvious care for her welfare shows through in a number of gestures, despite his whooping it up in Cape Town:

'I sent off three parcels for you of sweets, raisins, sultanas, cake mixture, butter, and fig jam (which I loved). The girl in the grocer's let us eat as much candied fruit as we liked. We went to the Assembly Hotel for lunch. We had soup, crayfish (a whole one, the size of a lobster), mixed grill, green peas, guavas and cream, by which time I was full up but the waiter brought a large piece of Roquefort cheese which I couldn't resist and then a huge packet of peaches, bananas, guavas, etc. which I made hay of, finally coffee and a cigar, the whole lot washed down with a bottle of beer, price 4/3d. Can you beat it? I had to undo my Sam Browne which wouldn't meet when we got up to go, and I was thankful to see Todhunter was in a worse state as he couldn't do up his top trouser buttons!'

They also ventured inland north to Groot Constantia, the oldest wine producing farm in South Africa, developed by Simon van der Stel in 1685. Then on his last day of shore leave in Cape Town, Pilk missed the route march as his feet were too blistered.

'Instead we set out in a seven-seater car to see the

countryside. We started off by shopping. I bought two lbs of slab chocolate, biscuits, etc., to fortify myself when the food on the boat wasn't up to much. We arrived in Stellenbosch and were deposited at the Grand Hotel where we had lunch — two large bottles of lager, soup, fish, a mixture of rice, mutton and vegetables, pineapple fritters, cheese, biscuits, fruit for 2/9d each and, as we had seconds of most, I thought it incredibly cheap. I bought more chocolate here. Nut milk slab at 2 per pound.'

After Stellenbosch, they ventured on to Paarl, another one of the main vine growing districts. They were by now informed that their original orders to go to Persia had been cancelled: 18 Division was to go to Bombay, where it would be held in readiness for either the Egyptian or the Far Eastern theatre of war.

'Originally we had been told that we were going to Egypt for four months' intensive training, after which we were to join 8 Army. But our orders, in view of Japan's entry into the war, were altered and we sailed for Bombay where we were to await further orders.'

Pilk, somewhere halfway across the Indian Ocean, still seemed more preoccupied by matters culinary:

'I was most amazed that I couldn't eat my crate of peaches that I bought at Cape Town fast enough and had to throw a number away which went bad. From Cape Town onwards we were escorted by one British cruiser, the Emerald. On 22 December, two days out from Bombay, our ship suddenly at 11 o'clock dropped out of the convoy, turned west and proceeded full speed to Mombassa.

Pilk had managed to send off a Cable & Wireless telegram to Phyllis at Woolacombe, which she received 23 December: 'ALL WELL AND SAFE. HAPPY CHRISTMAS AND NEW YEAR. ALL MY LOVE DEAREST.'

23.12.41

'Our voyage continues, uninterrupted by excitements. I do hope your telegram reaches me all right but, as we still don't know where we are going, I fear that it may be delayed. The food on board has been uninspiring but perhaps they will do something special for Christmas. The news generally seems satisfactory but it is sad to think of Penang in Japanese hands. I should imagine it will be some months before we can effectively counter Japanese initial gains, but I do hope that the powers which be don't lose sight of the real issue in Europe for, after all, when Hitler is downed, the defeat of Japan will only be a matter of days. Otherwise the war seems to be going well.' Penang was well-known to Pilk and Phyllis — many a home leave had they sailed the P&O liners from its evocative tropical port, past the glistening whitewashed minarets of the Eastern and Oriental Hotel, heading west toward England. The Japanese started broadcasting propaganda from the Penang wireless station, and within a few days dropped leaflets on Singapore calling for its surrender.

Mombassa, Kenya, was reached on Christmas Eve. This port call was shrouded in secrecy: 'An African port, very secret and no names allowed', was how Pilk had described it in one of his letters. Here the *Mount Vernon* refuelled and took on more supplies.

'Christmas Day was spent like any other day on board except that there were none of those interminable and useless inspections that the Army prides itself in. We all had a portion of a piece of turkey and some plum pudding to mark the occasion but, as over half the turkeys went bad, we did not get much! Since then we have been living on pineapples, bananas and oranges. Peter Westgate had to pay 2 each for his bananas because he said an officer should never bargain ...

The hotel proprietress, a South African woman, gave us a colossal tea of home-made cakes, bread and jam, and a huge dinner of seven courses and produced a large plum pudding for us with lashings of brandy sauce. Tea 1 and dinner 4. The second night she surpassed herself and produced a meal in our honour which almost beat me, but not quite, with huge pots of delicious home ground coffee.

We have been getting some queer dishes again. Pumpkin pie and a sort of mixture of blackcurrant and cherry, which I never get tired of. I have also a great liking for lime jelly. The game they give us is good, such as a whole pigeon or half a chicken, but sometimes they produce some weird bird that must be simply colossal as it is all one can do to get a leg on one's plate, but it suits me!'

After leaving Mombassa, Malaya was officially announced as their destination. Pilkington's extensive Malayan experience was duly tapped into:

'I was ordered to lecture all officers and troops, sometimes giving as many as four lectures a day, in which I described to them the country, the people and the jungle with its inhabitants, laying great stress on the fact that I considered the jungle was the only place to fight from, that it was not impenetrable or unfriendly, and that the side that learnt to make use of it would win.'

Pilk, as the defence officer under the American commander, was also ordered to arrange for the Ack-Ack defence. He mounted two Bofors and 36 Bren guns which, with the ship's guns, would offer solid fire power against any low-diving planes.

The last week in December 1941 saw the total routing of British forces from northern Malaya. The *Mount Vernon* continued steaming to 'a very secret naval base in the T. Group of islands' about 500 miles south of Colombo. This was Diego Garcia in the Indian Ocean.

'Two days out we intersected an outward-bound convoy from England which automatically split into two, some of the ships and escorting warships joining our ship and proceeding to Singapore via the T. Group base. This base was a large coral lagoon surrounded with palm trees. The only entrance — so narrow that a stone could be thrown on either shore from the deck when passing — was guarded by a boom. It could hold a whole fleet comfortably and, even if this isolated group of islands had been investigated by enemy planes, unless they actually flew over the anchorage, they would see nothing. At the base we were gradually joined by other ships carrying ammunition, troops and planes.'

They sailed the next morning for Singapore, being now escorted by the heavy cruiser HMS *Exeter* (which would be sunk by Japanese torpedoes in March 1942) as well as the cruiser HMS *Emerald*.

'My duties were almost entirely with my Ack-Ack guns. As we neared the Sumatra Straits, various units, battleships, cruisers, destroyers, and submarine chasers joined us from the Dutch Navy, while Catalina flying boats carried out aerial reconnaissance. All ships flew balloons or kites against suicide power-diving attempts to bomb our ships.'

They passed Java Head and the remains of Krakatau and, by nightfall, had cleared the Sunda Strait and were steaming through the shallow waters between Banka and Sumatra. In Malaya, the Japanese had a convincing win against 11 Indian Division at Slim River on 7 January and central Malaya fell.

11 January 1942 was a momentous day: the Japanese Imperial Army entered Kuala Lumpur, the federal capital of Malaya: 'We felt as though we had entered the crossroads of the central province of China,' said Lieutenant Colonel Tsuji Masanobu, Staff Officer in Charge of Operations, Japan Imperial Army. Duff Cooper, Resident Minister for Far Eastern Affairs, was recalled from Singapore to London by Churchill. And, unbeknown to Pilk, Phyllis gave birth to a son, Paul. That very day Pilk also penned what would be the last letter he would get through to home.

11.1.42

'It looks as if at last we are nearing the end of our voyage and we shall be thankful to get off this ship. The life has got very monotonous. It has been a very idle time. I suppose we shall miss the food off this boat terribly a few days after we have landed, but at the moment we are looking forward to something new. I must say they have contrived to feed us very well and how they succeed in carrying food for all the people on the boat for all this time is remarkable.'

By 12 January 1942 the Japanese had moved down into the state of Negri Sembilan, the very soil Pilk had tilled for over a third of his life.

'The *Mount Vernon* hove to at Singapore Naval Base at 15:00 hours on Tuesday 13 January in one of the heaviest tropical storms I have been in. For the last four hours British flying boats escorted us and the guns were continually manned. We later heard that flight after flight of Japanese bombers had been seeking out the convoy but that, owing to the thunderstorm, they had been unable to locate it.'

Japanese air force records confirm several aborted missions on those days.

'The ship was docked in the King George V graving dock so that, if hit, she wouldn't sink and could be quickly repaired. The rest of the convoy sailed on to the Singapore Harbour Board docks where they were bombed; one ship received a direct hit but was not scuttled.'

The Singapore Naval Base at Sembawang was completed at a cost of £63 million amid much fanfare in 1938. The largest British dockyard outside Britain, it was large enough to accommodate the entire British fleet. The *King George V* was one of two 50, 000-ton dry docks it boasted. The naval base was a prized possession in its own right, and — in a lovely case of circular reasoning — one of the key reasons why Singapore was so strategically important and to be vigorously defended in the first place. Besides, by 1930, 23 percent of Empire's trade passed through the Indian Ocean to Singapore. (In the case of Australia that figure was nearer 60 percent.)

The surroundings were little more than *atap*-hutted *kampongs* (villages) in among the plantings of the Nee Soon Rubber Estate, which, as foreign as it would have seemed to most of 53 Brigade, would have made Pilk feel strangely at home.

'I landed my company (C Company) at four o'clock and we were conveyed by Australian troop trucks to Tyersall Park Camp which is really part of the Botanical Gardens. It was still pouring and we were all soaked but when the rain stopped at about 17:30 we got the men into huts and they found a hot meal waiting for them.'

Tyersall Park Camp was of the hutted Indian Army pattern type with *atap* roofs. Very cool and very comfortable. It had been recently occupied by 1 Battalion Leicestershire Regiment who by now had been in the battle some weeks.

Tyersall Park had served as home to the fierce 2 Argylls, part of 12 Indian Infantry Brigade, since December 1939.

'What could be done to make us comfortable had been, only a meal for the officers had been forgotten but this we obtained by the simple expedient of breaking into a kitchen and helping ourselves.'

In the interests of secrecy, permission to write from Singapore was not granted (and would only be given to others 48 hours after Pilk had left for the 'firing line' up-country in Malaya). The telegram he was expecting to tell him of the good news of Paul's birth had not arrived.

14. Pilkington.—Our baby elephant, barges through everything. Unfortunately has been crocked for most of the season. Likes dough-nuts.

Pilk represented his state, Negri Sembilan, at rugby, soccer and hockey. And enjoyed a good party, too!

Pilk (left) at work as a rubber planter in Malaya. Later it was expected that they joined the FMS Volunteers and spend weekends training warcraft. He was friends of many of the officers of the Malay Regiment, formed at Port Dickson on the coast near Seremban.

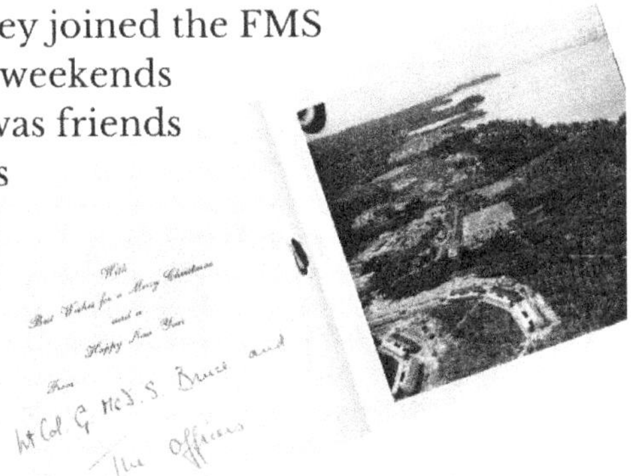

Pilk in an undated photo by Kunita, a Japanese photographer in Seremban (a popular pre-war cover for their operatives).

Below: Pilk and Phyllis married at St Mary's in Kuala Lumpur in 1932.

(Photos courtesy Pilkington family)

Above: The 6 Royal Norfolks embarked HMT *Duchess of Atholl* at Gourock, near Glasgow, on 28 October 1941.

Below: The whole of 53rd Infantry Brigade transferred to the *Mount Vernon* at Halifax.

2 COMING, READY OR NOT

Tyersall Park Camp, Singapore, 13–15 January 1942

'Late that night our kit began to arrive, most of it soaking wet as it had been dumped in the open in the pouring rain. I spent until 05:00 gradually getting sorted out. A lot of baggage and ammunition was still missing.

Next day, Thursday 14 was spent in drying clothing, unpacking, etc. Heavy kit began to arrive from the docks and transport from ordnance. We were told that we were to be given ten days to unpack and get sorted out as the Australian Brigade had been sent to Muar to hold the line and it was expected that they would be able to do this easily.'

The men had spent just six days on terra firma since October and had not been ashore since Cape Town, over a month before. Yet no leave into Singapore was allowed to officers or men. Pilk was well aware of the attractions of the city, having passed through many times over the years on official business, and on rugby tours when he'd come up against the Singapore Cricket Club team.

Singapore then was regarded as something of a Monte Carlo, a Buenos Aires of the East. In *Sinister Twilight*, Noel Barber called Singapore 'the last resort of yesterday in the world of tomorrow'. In this free-wheeling, cosmopolitan city, which boasted a certain savoir faire, people could kick up their heels in style. Coffee at Robinson's – whose building dominated the Raffles Place skyline – was fashionable. Their restaurant was even air-conditioned, an attraction in itself. Cold Storage offered fresh produce from all parts of the Empire. The latest books could be browsed at Kelly and Walsh, and ladies could indulge in a facial at Elizabeth Arden.

Then there were the *fin de siècle* charms of the Raffles Hotel (albeit out of bounds to all non-commissioned servicemen), Hotel d'Europe, Goodwood Park, and the ever-popular Seaview. The Raffles was still advertising its Dinner Dances ('from 7:30 pm till midnight') in the *Straits Times*, the daily masthead proclaiming

'Singapore must stand; it *shall* stand — H.E. The Governor'. The headline on 13 January trumpeted: **'JAPANESE SUFFER HUGE CASUALTIES'.**

The war was still deemed something distant, an irrelevant distraction to those in untouchable Singapore; a city staunchly in denial. Despite Singapore's population swelling from the usual 500, 000 to around one million with an influx of refugees from up-country, many foods remained unrationed, but even those that were, such as butter and meat, could be procured for the family in quantities three times the ration in England. The lights and rides of Happy World were a nocturnal magnet where its jazzy gaiety bloomed unabated.

But for Pilk's men, all the sights and joys of 'the impregnable fortress' remained off bounds allowing him, presciently, to dwell on their predicament.

'Our Brigade, which had just landed, was in no condition to go straight into action. Firstly, as regards my own Battalion there had been a change in commanding officers early in August, which had caused a good deal of upheaval as our previous CO had come from the Irish Guards and had been with us for over a year. All our training had been on Guards lines; in fact, so much so, that we were called the Norfolk Guards, a term of which we were very proud. But from now onwards we had to scrap all we had learnt and learn the regimental procedure and habits of the Norfolk Regiment.

In peacetime this would not have mattered much but, in war time, with a Battalion — nearly all territorials who wanted to get on with the war — little matters of dress, routine, etc. meant nothing, and unnecessary changes were always irksome.

Secondly, all the training of the Division had been for Libya and, for this purpose, the treeless districts of Scotland and Wales had been chosen. It was therefore difficult to expect men trained in such surroundings to plunge successfully into some of the thickest jungles in the world.

Thirdly, my battalion in mid-August had been out to Prescott outside Liverpool, where, as we were expected to be there for the winter, all war training ceased and everyone was put on to building a hutted camp under the Royal Engineers.

Fourthly, just before we were mobilised to go abroad there

had been an army comb out for technicians of all types with the result that my Battalion, like all others, lost some 20 percent of their best men and NCOs. Before leaving England, new drafts of conscripted men were sent us, all of whom were partially trained only, and the big gaps left among the NCOs had had to be filled with young and experienced men. In my own company I lost some forty men. The last replacements to reach me arrived just in time to embark and, when my company went into battle, I had literally never had my complete compliment of men on parade, and my platoon commanders had never been able to exercise their men in battle drill as a unit.

Fifthly, the brigade had been at sea eleven weeks in the most crowded conditions yet attempted when transporting men. Little or no attention could be given to physical fitness and weapon training as one's allotment of deck space only gave each man half an hour's training a day: fifteen minutes PT and fifteen minutes with their weapons.

Let it now be remembered that as a Division we were only considered half trained by the War Office and that, in order to bring us up to battle standard, we were to have *at least* four months intensive training in Palestine.

One could, therefore, hardly say that any unit of the Division was in a fit state for battle.

There is one point that should not be forgotten: the 18 Division was the most modernly equipped Division in the British Army, with firepower beyond anything that had previously been sent abroad, but, unfortunately, all its weapons were intended for open warfare and not close jungle fighting.

When we landed in Singapore the Malayan campaign had been in full swing and the British forces had retreated from Alor Star to Malacca, a distance of some four hundred miles in five weeks. Only British troops could have taken the hammering they did and still fall back in one unbroken line.

On our arrival we were told that a last stand was to be made on the Muar River which I knew to be theoretically and practically unsound, but my opinion naturally didn't count for much.'

Percival had been secretly planning the withdrawal to Singapore and the blowing of the Causeway since 30 December.

Australian Major General Gordon Bennett was vociferously opposed to what he termed the British 'withdrawal complex' and promoted an offensive rather than defensive posture. This found favour with General Archibald Wavell, supreme Commander of the Allies in this theatre, who insisted Percival change his plans. Westforce was thus established, with a new defence line — running from Segamat through Mount Ophir to Muar in Johor — a vital part of its strategy.

'The British air force in Malaya had almost entirely been shot out of the air, and those planes that still survived were no match for the Jap Navy Zeros. A small bomber force still remained intact but no turn for the better could be expected till the Hurricanes which our convoy had brought out could take the air.

Indeed, whenever a plane was spotted overhead it was assumed, generally rightly so, to be an enemy one. Up to 112 Japanese bombers would unleash their explosive fury on Singapore each day in January, which became known as "the month of the bombs".'

Fifty disassembled and crated Hurricanes arrived with the 53 Brigade convoy, but only 24 pilots were available. The planes had a brief impact against the Japanese, but within a fortnight nearly all were wiped out. Much of 53 Brigade's transport and bigger artillery pieces were on slower boats following with the balance of 18 Division.

'Our naval strength was so weak that they could not even prevent the Japanese light craft making landing after landing behind our forces on the coast.

The fall of Singapore at that time was certain. We all knew that we were not even a forlorn hope. We knew the battle was lost and that it was only a matter of time before we were captured or killed. We knew perfectly well that there were hardly any troops left to reinforce us and no one was ever taken in by the official pronouncements to the contrary.'

'Fortress Singapore' was one of history's greatest intended bluffs. Indeed, the myth of unassailable Singapore had been propagated by top military advisers, politicians, the media; and perpetuated around the bustling bar counters of the city. Bluster and bravado prevailed. 'To call it an island was a euphemistic misnomer. To call it a fortress was a hopeful misstatement of fact,' noted ex-18 Division Gunner A.G. Allbury in *Bamboo and Bushido*. 'Singapore in fact was a myth, and none knew it better than the Japanese.' Among the 3000 Japanese who called Singapore home were several fifth columnists, avidly observing developments, the lack of defences, and the traversable guns facing seawards to the south. (Six hundred Japanese families, along with the Consul-General, were recalled and evacuated in October 1941.)

In the early afternoon on 14 January, Wavell finally broke news to Churchill that there actually was no 'fortress' at Singapore. Churchill was reportedly horrified, calling it 'one of the greatest possible scandals that could possibly be exposed'. That morning Australian Major General Gordon Bennett formally assumed command of the Malayan front, encouraging the troops to dig in for a tide-turning consolidation in Johor. For Pilk:

'It was not till 15:00 that I was able to report that I was now prepared to go into action providing I was given twenty-four hours to clean weapons and sort out what would and would not be required.

At 16:00, while my company was on a route march in Singapore, I was hurriedly summoned to a Company Commanders' conference. Here we were told that the Japs had out-flanked the Australian positions on the Muar River by the simple expedient of walking through the jungle, but it was hoped to establish a new line further back.

Our battalion was to go to Ayer Hitam to cover the withdrawal if things went badly and, if all went well, we would be merely line- of-communication troops. It was found that the only two companies ready to move were C and D. I was put in charge of the convoy with orders to start at 10:00 next day.

By 19:00 my movement orders were completed and I handed over all final preparations to Captain Race and the platoon commanders while I arranged the collection and distribution of ammunition on a battle scale. It wasn't till 23:00 that the last details were completed and I was able to get some supper. Major Cubitt had very kindly sent me in a bottle of whisky so we had a cheerful evening and I turned in about 01:00 dead beat.

Next morning at 05:00 Carlton [Private C.H. Carlton, his

batman] called me, packed and sorted out my kit with great care. Fortunately, I insisted on filling my pack with five pounds of tobacco, which was as well as I never saw any of my kit again, but the tobacco lasted me five months.

Among other things I coated my feet with Calvert's soft carbolic soap and, although I never had my boots off for a week, my feet came out spotlessly clean and unblistered.

At 09:00 I paraded my company, inspected arms and ammunition, made last arrangements for feeding and iron rations. At 09:45 I embussed the Company; D Company having embussed they joined the convoy and we moved off at 10:00 to the second.'

(On 11 February, less than a month after Pilk passed through, incendiary bombs were dropped by the Japanese air force onto Tyersall Park camp, by then home to the Gordon Highlanders who had already been bombed out of their Birdwood Camp base. Seven hundred wounded Indian soldiers at Tyersall's Indian Base Hospital were killed as the thatched *atap* huts went up in a roaring inferno.)

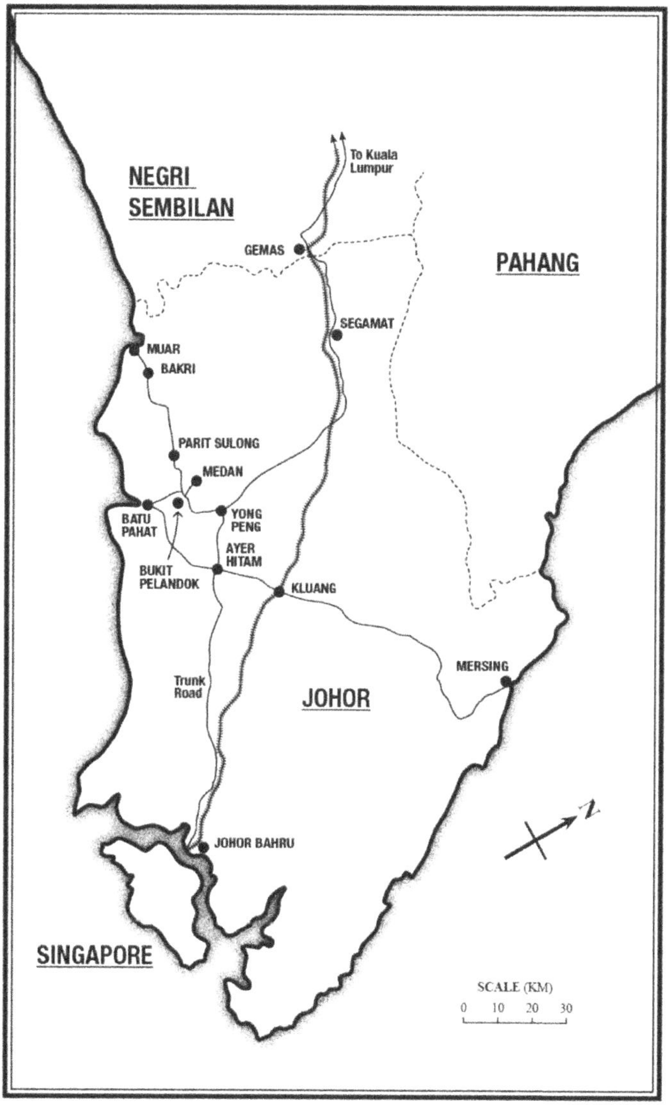
To Kuala Lumpur
NEGRI SEMBILAN
GEMAS
PAHANG
SEGAMAT
MUAR
BAKRI
PARIT SULONG
MEDAN
BATU PAHAT
YONG PENG
AYER HITAM
BUKIT PELANDOK
KLUANG
MERSING
Trunk Road
JOHOR
JOHOR BAHRU
N
SINGAPORE
SCALE (KM)
0 10 20 30

View of the battlefield from Muar across to Bukit Pelandok, rising rear left. (AWM 117523).
Below: The blown bridge,Yong Peng. (AWM 117524)

3 INTO THE FIRING LINE

Johor, Malaya, 16–23 January 1942

'We crossed the Johor causeway about 12 o'clock. It was very noticeable that no attempt had been made even at that date to defend it or demolish it.'

Brooke-Popham's prime objective was defence of the naval base, to the east of the Causeway. Hence beach defences along this north and the north-western shore of Singapore were considered irrelevant and never materialised. Brigadier Ivan Simson had been dispatched from London specifically to address construction of defences in Malaya and 'Fortress Singapore' in August 1941. Percival seemed resentful of his role, and Simson's exhaustive surveys and resultant recommendations were consummately ignored. Percival even famously rebuked his pleas with: 'Defences are bad for morale, for both troops and civilians.' Meanwhile, Pilk was passing through Johor Baru, 'where we found the Australian Red Cross unit evacuating their base hospital to Singapore. They had a cheerful lot of sisters who threw armfuls of cigarettes into the trucks as we passed. About 1 o'clock I halted the convoy in a rubber estate for an hour's rest and we watched an endless stream of refugees – both European and native – going by, all with cars piled high with their household goods.'

When the crack Japanese 5 Imperial Guards Division under Lieutenant General Nishimura Takuma had cut the vital Muar–Yong Peng Road in Johor State strategic Yong Peng was exposed to a flanking action. This was felt acutely at Bennett's Westforce HQ at Segamat, 112 kilometres north of Yong Peng. He and his men were in danger of being isolated and overrun. Bennett travelled down that day to his rear HQ, six kilometres east of Yong Peng, to meet Percival at noon. Between withdrawing and stabilising, they decided to shore up the Muar front at Bakri to where 45 Indian Brigade had retreated.

'At two o'clock we started north again and from now onwards were escorted by Japanese bombers who, however, did not molest us. At about 3:30 I halted half a mile south of Ayer Hitam crossroads, where I heard that the Brigadier had been bombed while, against all the rules, holding a conference on the road. Here a dispatch rider brought me orders to proceed to the 12th mile Yong Peng-Muar road with OC D Company, the rest of the convoy to follow under second in command. From here onwards the roads were clear of all traffic.

About 5 o'clock I met my CO at the road junction leading to the Japanese-owned tin mines of Medan. Here he told us that the Australian and Indian brigades from the Muar River would withdraw through us that night. In the meantime we were to cover the road approaches in support of a composite British battalion of the Loyals that would hold the main ridge dominating this piece of country. I was allocated an area four hundred yards down the Medan road while D Company took over the area west of the T-junction on the Muar Road. By the time my reconnaissance was finished, my company had arrived, debussed and was immediately put to digging in. By 6:30 tolerable defences were dug and manned, and I ordered the evening meal, which we had brought in hay boxes, to be served.

In the meantime I went down to the village of Medan, some two miles away, where all was quiet and the police were still in charge. On my return a Jap fifth columnist sharp-shooter had a crack at my car but a burst from a tommy gun stopped that nonsense. On my return to Company headquarters I was met by my CO. It was now nearly dark and he said that the battalion of Loyals had not arrived and therefore C and D companies must move forward to cover the defile through the hills. All our carefully worked-out positions had to be abandoned and, in pitch darkness, I had to re-deploy my company in a country where I had only the crudest map to guide me.

By 9 pm new positions were occupied, more as an ambush than a defence as, at that time, all Jap attacks had been made in the early dawn in mass formation down the roads. At 10 o'clock the police at Medan produced forty confiscated bicycles so, in conjunction with the CO and OC 'D' Company, bicycle patrols were organised to cover the area between Batu Pahat Ferry and the bridge at the 20th milestone.

Double sentries having been posted and the platoon commanders given their orders, I turned in for two hours' sleep at midnight, being woken at 2 am by Captain Race, my second in command, and I took over till 4 am. My CSM carried on till 5:30 when I ordered the stand to.

Daylight showed that I had been allotted a very steep bare hill with a forward slope, quite impossible to defend. Further it was a landmark for all artillery for miles around. I therefore made my dispositions on the edge of the surrounding jungle. These were approved by my CO who, at the same time, told me that the rest of the battalion had been ordered up at midnight and should arrive in support about dawn.'

The 6 Norfolks had moved onto Bukit Pelandok, 10 kilometres west of Yong Peng. A defile runs between 500-metre Bukit Pelandok and Bukit Belah, covering the all-important triangular junction connecting Yong Peng to Parit Sulong and Batu Pahat. Eight kilometres further down, a Norfolk platoon was defending a bridge at Parit Sulong. The Japanese were stealthily advancing north from the direction of Batu Pahat toward Pelandok, and south-east from Muar to Bakri. Their aim was singularly to surround, and annihilate all Allies there. Yamashita described the battle at Bakri 'the most savage' of all the Malayan campaign. According to Pilk:

'At one o'clock the Divisional Commander appeared and told us that the Australian and 45 Indian Brigade had decided not to withdraw but had taken up new positions about 25 miles up the road. At the same time he advised me to re-dispose my positions so as to cover the open hill, which we all knew to be madness. He agreed but said it was essential that this be done. I reallocated positions on the hill itself, which we dug and completed by 3 o'clock.

At 3 o'clock I was ordered to construct a roadblock and was supplied with 24 one-ton concrete blocks which were dumped on the side of the road. These took over three hours to get into position. An anti-tank gun arrived which I placed on the hillside covering the block. I was told to allow the RE unit of the Johor Volunteers to withdraw from the Batu Pahat Ferry. This they did, passing through my lines about 4 o'clock. They were, however, supposed to have demolished – and had reported to Brigade that they had done so –

the temporary bridge they had constructed across the Sungei Kori River and the old Ferry. In order to satisfy myself I took a heavily armed patrol down the road and found that their demolitions had been very incomplete and the river could very easily be crossed. They had also abandoned their water cart, worth its weight in gold.

I reported to the Brigadier who sent down our own sappers to complete the demolition; they were sniped while at work. Griffiths, our motor transport officer, collected the water tank which was used by the Battalion.

By now heavily armed patrols from 'A' and 'B' Companies were covering the roads in front of us, a Japanese outflanking force being reported about twenty miles away. Night patrols and double sentries were arranged and the night passed without incident.

Early the next morning, Sunday, a Japanese recce plane came over but did not molest us. I ordered all slit trenches to be deepened 12 inches to give added protection against bombing.'

Bennett was now realising the full extent of the Japanese threat: not 500 as he'd been told by British Intelligence; rather, an entire division of Japanese Imperial Guards was deployed in and around Muar. They had approximately 10, 000 troops, supported by scores of tanks – dismissed by the British planners as unfeasible in the jungles of Malaya. He decided it was prudent to withdraw Westforce behind the Segamat River, ahead of a broader withdrawal south.

'The Brigadier and CO visited me during the morning and informed me that the troops in front would be withdrawing and that a promise had been extracted from the Divisional Commander that we should be withdrawn from our untenable position as soon as possible.

At 5:30 the evening meal arrived by truck in spite of the fact that it had been ordered for eight o'clock when it would be dark.

The mobile van came to the front lines bearing mainly dry rations, such as dried milk and dried fish, but the main bulk was that staple army ration, bully beef.

As luck would have it a formation of nine Jap bombers flew over as the meal was being unloaded and promptly started bombing us. In all, in the ten minutes, they dropped some thirty-six heavy bombs in an area about twice the size of Barricane Beach [A small exotic shell beach cove tucked away in Woolacombe, Devon]. I lost four killed and eighteen wounded, and only the fortunate deepening of the slit trenches saved the rest. My Ack-Ack Bren gunners were magnificent and kept their guns in action throughout the bombing.

As soon as the wounded had been evacuated I examined the damage and found that nearly all our positions, so carefully dug and camouflaged, had been literally blown out of existence, and my Company HQ was no longer tenable. It was too late that night to do more than re-allocate those positions left intact and to move Company HQ back down the road to a swamp some fifty yards away.

The night passed quietly for us. Much to my annoyance I was woken up by an urgent telephone message enquiring how many cigarettes and how many tobacco rations I required. Needless to say I never saw either.

Next morning, before dawn, I buried those killed in a communal grave.

The new positions were occupied but, with the depleted numbers, could not be fully manned. Most of the morning I had to act as interpreter and, to my eternal satisfaction, was able to save a number of perfectly innocent Chinese from being shot as fifth columnists.'

Pilk found himself less than 100 kilometres from where he'd lived harmoniously on the rubber estates with Chinese friends and co- workers, as well as Malays and Tamil plantation workers (Peninsular Malaya being home to 750, 000 Indians in 1941). A full 75 percent of rubber plantations were in the hands of European-owned estates such as Dunlop and Guthrie's. Pre-war, the Federated Malay States (covering Negri Sembilan, Selangor, Pahang, and Perak) were answerable to a federal government in Kuala Lumpur. Sultans ran these states independently, with 'assistance' from a British Resident. The Unfederated States similarly were run by Sultans and their British Adviser 'assistants', but did not report in to the KL government.

Because of this unwieldy structure, communication — let alone change — was nigh on impossible to facilitate efficiently and effectively. But if it ain't broke ... at no point from 1895 to 1941 did Malaya require any financial help from outside. It was the model colonial economy, paying its way to healthy surpluses through tin production and the 2 million acres of rubber plantation under cultivation as at 1938.

Japanese, too, were no strangers to this land given their investments totalled 85 million Yen. Other than mine owners, many were chemists, barbers and photographers in the cities and towns.

Japan had been planning this campaign for thirty years and the country was riddled with their agents and fifth columnists. Further, throughout this battle zone they had smuggled in, during peacetime, masses of rifle and mortar ammunition, all of which was waiting for them as they advanced, in secret dumps.

'The Japanese soldiers were trained in jungle warfare: they were superb if ignorant fighters and they had been taught to live off the land. Supply questions, that would have halted an English army, for them did not exist; a little rice and boots was all they wanted. The only difficulty they experienced was getting their rice cooked.

At 12 o'clock the Japs were reported four miles away and I gave orders that no more transport was to go forward past my roadblock. Incidentally, this saved an Australian transport column of some 30 trucks from falling into the enemy's lines. Reports of Japs in the jungle and up the trees invariably proved to be monkeys.

At 12:15 I was given orders by my CO to send out a patrol of an officer and twelve men to cover the jungle to my left. This patrol was never seen again. At 12:30 I was ordered to send an armed patrol of an officer and a full platoon to cover the Chinese market gardens in front. I said it was impossible, but was told it was a direct order from the Brigadier. Instead I could, in view of my casualties, cut down the patrol to twenty men. This patrol left at 1 o'clock under Lieutenant Parker.

I then examined Lieutenant Nutt's position on my left flank and those on the hill and personally watched sentries placed well out in the jungle to guard against surprise. We had no wire to prevent our positions being rushed as none had arrived.

On Monday at 2 o'clock the Japs put in a surprise attack from the jungle on the left. The first intimation I had that they were there was seeing one of them standing out in the open below the hill. The attack soon developed into a full-scale operation in which my depleted platoon held their own until the Japs started firing down on them from the tops of the surrounding trees, for which purpose they carried climbing irons.

By three o'clock Lieutenant Nutt's position in the jungle had held fast but the hill had been carried, the survivors making a getaway down the road. This left my Company HQ in the air, the anti-tank gunners who covered my rear having withdrawn without orders.'

The strategic Bukit Pelandok had been taken. Orders were immediately issued to recapture it, given its strategic importance in governing the Muar-Yong Peng Road.

'I was still in telephonic communication with Brigade HQ and, with them, arranged for a counter attack to recapture the hill, this to be carried out by a company of Loyals. They never appeared.

The Loyals later stated they had lost their way while trying to strike the hill from the flank through the jungle.

By 3:30 the Jap mortars had found the range of my company positions and, for the next two hours, pumped shells on to me at an average rate of four a minute. Two out of my eleven men were killed. My telephone wire was blown to bits but Captain Race crawled out and joined the broken ends.

By now the Japs had planted machine guns on the captured hill which would make anyone withdrawing down the road come under very heavy fire. I was told that our carriers, which had been out in front, would make a dash to break through the cordon. Some twenty minutes later we heard them coming. They were magnificently driven and rounded a right-angled corner at full speed. As they did this I fired at the Jap machine gunner and saw him roll over and over down the hill into the road. The only Jap I know I killed. Race shot the Jap officer — who foolishly stood up on the horizon — and the carriers were through without a casualty.

After this, the mortaring of my position increased in fury and I was ordered by my CO to withdraw at nightfall. The battle was raging all round me and I was completely cut off. I was, however, able to shout to Captain Westgate of D Company to cover his left flank as I could do no more for him. Strangely enough this company was not attacked till they had been isolated two days, by which time they were able to beat a retreat through the jungle.

At dusk I buried our dead, destroyed our stores, but took out all our fighting weapons. I led the men through the jungle in a wide circle and it was not till we came out on to the open ground between us and Brigade HQ that we were fired on. I found a culvert down which we crawled, the Jap bullets going harmlessly over our heads. Our final dash to safety was covered by Lieutenant Evans and Lieutenant Goddard with the Carrier Platoon. Once round the bend of the road I met the Brigadier to whom I made my report. With him was a colonel of an Indian Ghurkha Regiment which was preparing to put in a counter attack next morning. He was killed next day.

My report that over a thousand Japs were now occupying the ridge was scoffed at, which was a pity as it would have saved many lives.

At midnight I got to sleep and slept till dawn when the attack by the Indians was put in. Suffering heavy casualties, they captured the ridge only to be driven back by the Japs a little later. They contacted D Company who, unfortunately, never quite realised the position and a general fight took place between Indians, English and Japs and, by the time it was sorted out, it was too late for D Company to withdraw and their commander Captain P.M. Westgate was killed in this action.'

On 20 January, Westforce HQ was relocated south to Yong Peng, Bennett ordering a retreat to this town. Yamashita's HQ was by now firmly established in Kuala Lumpur.

'On this day, Tuesday, the battle became general (our artillery having come up), and more Indian troops, the remains of the Battalion's two A and B Companies and HQ Company still intact, withdrew to the police station about a mile down the road. Here I was ordered to take over the defence of the Battalion and, for most of the time, acted as second in command.

We held this position all Tuesday, Wednesday and half of Thursday. We were continuously bombed from the air and machine- gunned, but, as by now during the day time I was allowed to keep all troops under cover in the jungle, we had no casualties except when once again the ration truck was bombed and machine-gunned in broad daylight, two men being slightly wounded. After this all supplies came out by night.'

On 21 January, Bennett was attempting to get the 2 Loyals through to Parit Sulong to rescue Lieutenant Colonel C.G.W. Anderson and his besieged 2/19 Australians. They, along with 2/29 troops also in Anderson Force, had been sandwiched by Japanese advancing south- east from Muar and north-west from Batu Pahat. The Japanese now held the convex concrete bridge after a 6 Norfolks platoon, having not been re-supplied for two days, abandoned it fearing they had been cut off. The British attack to retake the bridge, originally slated for early that afternoon, was postponed three times before being cancelled, with no explanation ever forthcoming.

'During these two days I had to send back at least six men whose nerves had been unable to stand the sleepless nights and incessant air attacks. It was always a great relief when the Jap bomber formations, generally 27 at a time, passed us by and we heard them bombing someone else. Once while walking down the road a Jap fighter passed over my head, the pilot very kindly throwing a couple of hand grenades at me which burst harmlessly in the swamp.

It was terrible to see the fear the animals had of airplanes. Squirrels and birds would rush screaming all over the trees.

On Thursday morning Lieutenant Colonel Lywood, suffering from acute malaria, had to be sent back to rest with the transport column some ten miles back. Major Cubitt took command and I took over second in command. Much to my joy Lieutenant Nutt, having evacuated his positions at midnight as ordered by me, turned up with the twelve survivors of his platoon having carried their wounded twelve miles through the jungle in Batu Pahat, guided by his compass and the Southern Cross. A magnificent effort; but then he was my best officer.

On Thursday at midday we were told to fall back another two miles down the road to cover a general withdrawal which was to take place next day. Our new positions were in open country, flanked by jungle and covered in *lalang* grass four feet high which, if it did prevent the Japs seeing us, also prevented us firing without standing up.

Here we received a little dannert wire which I had strung in the jungle, preventing the Japs rushing this position when the final withdrawal did take place. I also had the opportunity of sending Carlton to Brigade HQ with some Australian soldiers that had got through. I was not to see him again for six months.

We got a really large and hot meal that night and nearly everyone was able to get six hours' sleep.'

On that day, 145 seriously wounded (many immobile in hospital trucks), had been bayoneted, machine-gunned while tethered, and/ or burnt alive at Parit Sulong after Anderson's failed negotiation with the Japanese to let the wounded out. Three miraculously survived, but only one — Ben Hackney — would later tell the tale of one of the war's worst atrocities. For his part, Anderson received the only VC in the Malayan campaign. Hackney received no decoration.

Singapore Free Press, Friday 23 January: **'STRONG COUNTER- DRIVE IN WESTERN MALAYA!'**

With Percival increasingly desperate, Gordon Bennett moved his HQ south from Yong Peng to Rengam and sent out orders to hold the Batu Pahat–Ayer Hitam–Kluang–Jamaluang line. No further withdrawal, Percival ordered Bennett, was to be made without his personal permission. At midnight the Yong Peng bridge was blown by the 2/30 Australians as the last Allied troops fell back to the new line. Just 550 Australians and 400 Indian soldiers of the original 4000 defenders of Bakri reached Yong Peng.

'The withdrawal of all British forces to Ayer Hitam began. It was no longer possible to hold the positions in case the Australian and Indian Brigades could fight their way out. We were heavily outnumbered and had no air support, while enemy artillery was getting stronger and nearly the full force of their bomber command was thrown against us.

It should be noted that the Jap troops brought up to clear these positions — which would leave the road to Singapore open, and at the same time cut off our troops withdrawing down the road and railway from Tampin — were the Japanese Imperial Guards: fearless, skillful and magnificent fighters.'

Distinctive in their olive green uniforms, the Imperial Guards were tall and physically robust, unlike Allied propaganda would have it. (It was true, however, that the average Japanese man was then six or seven centimetres shorter than his Western counterpart.) The Sino War had equipped them with a doctrine to accept natural obstacles to avoid contact with the enemy. Hence the Allies were continually surprised by their overland advances through supposedly impenetrable jungle, and were unable to bring cohesive force to bear on the numerically- inferior Japanese. According to Lieutenant Colonel Tsuji, Chief of Operations and Planning, Japanese 25 Army: 'Thanks to Britain's dear money spent on the excellent paved roads, and to the cheap Japanese bicycles, the assault on Malaya was easy.'

'By 12 o'clock the withdrawal was in full swing. By 2 o'clock our forward line had been abandoned. By 3 o'clock all bridges had been demolished forward of the causeway, so called as the road here was floated for a mile over swamp.

At 3:30 the Japs sent light tanks to clear the roadblock, but one of our anti-tank guns got in a direct hit with its first shot which knocked the Jap tank out of action and slewed it across the road, preventing any further tanks passing. The Brigadier, who had withdrawn to our position, however, was taking no chances and blew road craters in the defile running through our position. This came as a complete surprise to us as we had only a couple of minutes' warning. A huge crater was made which would stop any tank, but we had all been covered with mud, and many of us badly cut and bruised by lumps of flying mud and rock. It also cut off the withdrawal of the anti-tank guns which had to be abandoned. These became a nice little target for the Jap mortars.

By four o'clock we might be said to be in contact with the enemy although, owing to the high grass, we could not see them but could only fire by sound. Here, owing to the impossibility of getting it back, I ordered our 200 rounds of 3" mortar ammunition to be fired on a compass bearing to cover the road. That mortar fire may have killed no Japs but it certainly slowed up their attack and put new heart into my men.

About 5 o'clock Major Cubitt and I made final arrangements for our own withdrawal at 6 o'clock. Around 5:30 a mortar shell burst about ten yards from the slit trench in which I was. The concussion ruptured my right eardrum and seriously damaged my left, but hearing in the left ear returned and was quite good.

At the same time, the Japs were reported assembling in a hollow about 300 yards away and I went round making final dispositions to beat off this attack. While doing so I spotted a Jap sniper up a tree at whom I had a shot with my sergeant's rifle. I think I hit him as I saw him sag over a branch.

But, unfortunately, standing up to take this shot over the grass exposed me to a Jap sniper in the rear who put a heavy bullet through my shoulder from the back.

The shock spun me round, and for a few minutes I thought I had been hit through the chest — all the pain was in the middle of my back where the muscles had been wrenched out with the force of the shot. My CSM Kelf dragged me to the rear where Major Cubitt ordered me back at once.

For the next 20 minutes, with my arm slung in the braces of my equipment, I felt strangely on top of the world and had no pain although I was, unknown to myself, bleeding dangerously down my back. I stopped for a few minutes to enquire the proper destination of the anti-tank guns and wagons when I was joined by John Stratford and another of my sergeants, CSM Twiddy, who had been detailed by Major Cubitt to help me back.

We started off down the road which was being heavily shelled, so much so that I ordered Captain Stratford and Sergeant Kelf to keep well off the road and meet me when we had rounded the corner down the road. CSM Twiddy almost literally carried me this distance as I was beginning to feel the loss of blood, and we safely negotiated the next 3/4 mile through the anti-tank mines.

By now the pain was almost unendurable and, unfortunately, the Morphia I carried for myself and my men for such an emergency was in the breast pocket underneath my web equipment. Attempts to try and get at it, which meant moving the arm, had to be abandoned as I could not stand the pain.

About a mile down the road was a heavy wooden bridge over a deep muddy river some thirty feet wide. The bridge had been blown and all that was left was one long round tree trunk all slippery with mud. Under normal conditions it would not have been easy, but under the conditions no one could help me. I miraculously crossed the bridge without falling, and lost consciousness on the far bank.

Here Stratford and the two sergeants got hold of an abandoned bicycle on which they wheeled me for half a mile till we reached the next bridge. By this time I had recovered consciousness again.

I was transferred from the pushbike to a stretcher under the care of the RSM, and Corporal Cole of the RAMC tried vainly to stop the bleeding.

By now I was drenched with blood. Fortunately a despatch rider had been sent off to collect a spare ambulance which Captain Griffiths, our transport officer, had collected. The Brigadier and John Stratford wished me good luck and I was loaded in. I never saw Stratford again.

[Stratford, who had been Pilk's cabin-mate on the *Mount Vernon*, was killed on 12 February in defensive action near the Singapore Naval Base.]

Of the journey to the advanced dressing station I remember little; it was all a ghastly nightmare of pains that wouldn't stop. I can remember the medical orderly with me speaking to Griffiths and asking him to go faster; and Griffiths replying that he didn't dare as he wouldn't be able to dodge the shell holes. I can also remember swallowing a whole water bottle (three pints) of water without stopping.

Before crossing a bridge I can remember the ambulance stopping and Captain Johnson, our MO, looking at me and telling Griffiths to rush me on to the advanced dressing station. Eventually, after what seemed hours but was really twenty minutes, we reached it. I heard the cheerful voice of Captain MacDonald, one of our Field Ambulance MOs, saying, 'Hello, Pilk, here you are at last. We have been waiting for you.' I was lifted out and carried inside the tent where I was immediately given a full grain of Morphia. Within a minute I felt a different person, as though I was acting in a dream.

They cut away my shorts and searched my tummy as they were sure I must have been hit there as well to produce all that blood. In the meantime the doctor and orderlies were trying to stop the bleeding, in which they were only partially successful and I can still remember the MO ordering an ambulance to be ready to take me immediately to the Casualty Clearing Station.

My wound dressed, they cut off my boots and stockings, but didn't dare touch my equipment for fear of aggravating the bleeding. I was then wrapped in blankets, surrounded with hot water bottles, and given as much hot sweet tea as I could drink. I lowered three mugs in quick succession. I was reloaded into an ambulance with a medical orderly to look after me, together with a three-pint flask of water. MacDonald found my pipe and tobacco, which he lit for me.

The journey to the CCS, about 15 miles, passed in a pleasant dream. My pipe smoked well, I polished off the three pints of water and was violently sick, which didn't worry me. I felt no pain except a drowsy sort of itch. I remember the ambulance stopping about half way and a doctor examining the dressing and telling the driver to push on.

It must have been about 8 pm that the ambulance drove into a rubber factory which was being used as the CCS. Here I was unloaded and a surgeon came over and examined my wound, whereupon I was taken into the annex to the operating theatre. My stretcher was placed on trestles, and I was put under the personal care of an Australian Red Cross staff sergeant who never left my side and did everything for me.

For what appeared ages, I waited while I was given injection after injection of adrenaline and something else to buck up my heart. I was very sick several times. Eventually the surgeon appeared satisfied, when a portable X-ray apparatus was whisked up. Almost as soon as the photographs were taken they were developed and the surgeon told me that, although the wound was a serious one, he hoped to save the arm. I was then carried into the operating tent. The anaesthetist came up, told me to breathe deeply and, with the second breath, I was unconscious.

Some hours later when I came out of the anaesthetic I found myself in a real bed in a big marquee with a doctor by me. What he said to me I don't know, but I was given a powerful sleeping draught and didn't wake up till about six next morning. I found I had been sheathed in plaster of Paris from my neck to my thighs and could only just move the unwounded arm. An orderly by my bed called the doctor who satisfied me that my arm hadn't been amputated! I was quite convinced it had as, not only couldn't I see it, but I couldn't even feel it.

Later I was given a large mug of cocoa and brandy, and I was then transferred to a stretcher and loaded into another ambulance. Just before I left I was given more Morphia and half a glass of neat brandy, with the result that the ensuing journey to the Alexandra

Base Hospital was like riding on a cloud. My orderly lit cigarettes for me. Some more brandy and all was well with the world.

When I was taken out of the ambulance I was merrily drunk.'

4 BLEEDING SLAUGHTERHOUSE

Alexandra Military Hospital, Singapore, 24 January–26 February 1942

The languid cantonment of Alexandra used to be little more than palm trees and other tall, thickly-headed trees, to the north, and a train track that ran roughly parallel to another track, Ayer Rajah Road. Along Alexandra Road were small Chinese kampongs, bamboo and *atap*-style huts and a shop. Fruit stalls sold pineapples and papaya freshly picked from the adjacent fertile fields. On the western side of Alexandra Road were more coconut palms plus old rubber plantations.

But with the increase of garrisoned pre-war troops putting pressure on the Tanglin Barracks' 240-bed hospital, it was deemed a newer, larger facility was required. And so a twelve-hectare site in Alexandra was designated, and Major J.W. Colbran of the Royal Engineers set about designing the 356-bed hospital and ancillary buildings; for Europeans only.

Construction began in 1938 and the three-storey red-roofed white building was long and imposing, towering above the surrounding *lalang* grass and shrubbery. Upon its completion in late 1939, Colbran and his team were justifiably proud that they had created what the *Straits Times* considered 'the most up-to-date and one of the largest military hospitals outside of Great Britain'. But it wasn't long before this medical masterpiece was considered to be too small. With the looming spectre of war, in April 1941 GOC Malaya sought permission of the War Office in London to expand capacity to 600 beds. GOC estimated that there was a need for 4684 more beds for Europeans ('excluding Australians') and Non-Europeans in Singapore. However, it was deemed unfeasible to expand it to any more than a 450-bed capacity, and that only by cramming more beds into each ward and using some of the generous verandah spaces.

Meanwhile, by the last two weeks of January 1942 the rampant Japanese had cleared Pilk's old stomping ground of Negri

Sembilan, and even Mersing down on the south-east coast. The Allied casualties were unimaginably high with 13, 000 British soldiers already dead in the battle for Malaya, and all serious cases — such as Pilk — ferried to the Alex.

'The officers' wards were at the top of the building on the third floor, open on both sides with broad balconies and very cool all day. I was whisked off in a lift and met at the top by a cheerful Scot sister who conducted my stretcher, feet foremost to her ward, which is no way to treat a living being! Four orderlies and three sisters rallied round and I was lifted most gently into bed, but the movement was enough to start the pain again but more Morphia quietened it down.

My shoulder swelled up inside its plaster case to such an extent that the flesh burst in places. For the three weeks the sisters were with us I was never allowed to be touched except by them or the experienced first-class nursing orderlies told off for bad cases. Anything I wanted or asked for was brought to me and the VADs [members of the Volunteer Aid Detachment] were most charming and helpful in such little matters as lighting my pipe, cutting up my meals and, during the first few days, feeding me. In order that no one should jar my bed, a veritable barricade was built round it. I was thoroughly spoiled and enjoyed it: clean sheets three times a day. Powdered and washed like a baby morning, evening and night, and all the choicest titbits they could give me.

I was so weak that the slightest effort on my part brought me out into a cold sweat and uncontrollable tears. For the most part I lived on M&B and Dr Clark's blood pills, liver, chicken and milk, with as much beer and whisky as I wanted. The latter always amused me as the sister would come round, carefully measure out the prescribed amount, and then come back half an hour later and leave a bottle and some soda by my bed in order to get me through the night.

At 10 pm, sister would come and powder me up, give me a huge mug of Ovaltine followed by a strong sleeping draught, when I would sleep till four or five in the morning, then I would be given cups of tea.

On Monday January 26 I was wheeled into the ear specialist's theatre where he performed wonders on my right ear, eventually sealing it with some powder. He was the soul of kindness and always looked me up twice a day. At this time the pain from my ear was a strong rival to my shoulder.

On Tuesday 27 I was taken down to the X-ray room and in the afternoon blood examinations were taken. I should have been dying as my blood count was only forty and the danger point was sixty. After this I was made to eat masses of liver; the head sister, a Miss Currie, sat on my bed to see that I did so.'

Singapore Free Press, Wednesday 28 January: **'Will fight to last inch in Johor — Churchill', 'Malaya greatly reinforced last week'**; Thursday 29 January: **'Evacuation of North Coastal Area. Precautionary Move Taken in Singapore.'** A wonderfully optimistic classified ad also appeared in the publication that day: 'For Sale: European Guest House. In select, quiet non-military area. Now absolutely full, good proposition, would also consider lease for fixed period. Reply with bank reference to PO Box 309, Straits Times.'

'On Thursday 29, with my temperature still rising, it was decided to operate again and at six o'clock in the evening I was collected by two very clumsy orderlies, who hurt me so much in moving me that I was not able to be operated on for another hour. Then instead of a general anaesthetic they had to operate by Morphia alone. Although I felt very little, that half hour was a nightmare as I had to be supported in a sitting position by two sisters, while the plaster was cut away, the wound cleaned and bandaged. On my return journey I was accompanied by the theatre sister and this time I was handled so gently that I hardly felt a thing.

A new bed had been prepared for me, outside by the doors so that a continuous draught blew over me. Up to now I had had no new dressings but for some weeks my wound was dressed every morning under the direct supervision of the doctor and Sister Currie with two others to help. Orderlies were discarded as they couldn't stand the ordeal. Always afterwards I was given a large glass of brandy.

The Surgeon, a Major Webster, was kindness itself, and I cannot say how much I owe to him. The sisters were one and all super- humanly kind, and particularly Sister Currie and Sister Bowers and the night sisters. The VADs, amongst whom was Betty

Wiseman, loaded me with fruit and anything they could find. There was one VAD who had been a children's nurse all her life and, not having any children to look after, took me over.'

The balance of Major General Beckwith-Smith's British 18 Division, consisting of 54 and 55 Brigades, landed in Singapore together with all divisional troops — four days after the Allies' decision to forego mainland Malaya.

'When they were landed the fall of Singapore was a foregone conclusion, and I am doubtful if they were able to postpone the capitulation by more than two or three days.

My battalion, after withdrawing from Yong Peng to Ayer Hitam, were told to help hold a line on the Ayer Hitam–Batu Pahat Road. Here they were ambushed, cut off and eventually had to cut their way through the swamp and jungle to the coast, where they were picked up by the Navy and evacuated by destroyers. Lieutenant Nutt made a gallant attempt to extricate some men cut off, with carriers. He succeeded only to lose his own life. He was recommended for a posthumous VC.

During this action our first reinforcements, consisting of some seven officers and 80 other ranks, were ambushed at Senggarang village. They never stood a dog's chance and were wiped out to a man, the Japanese bayoneting those that surrendered after which they piled the bodies, some still living, in a heap, poured petrol over them and set fire to them.'

Singapore Free Press, Friday 30 January: **'Curfew for Singapore from 9pm to 5am. Order comes in force tonight'**, **'Fierce fighting in Johor: Aussies score success.'** Elsewhere in the paper, the Raffles Hotel was still advertising — 'Tonight dance 8pm to midnight' — as were the Adelphi and Seaview Hotels with their 'orchestral selections and popular concerts'.

On 30 January Pilk sent a telegram birthday greeting to Phyllis, whose birthday was on 1 February, reading: 'MANY HAPPY RETURNS. ALL WELL AND SAFE, LOVE.' She would not receive it for 3.5 years. Meanwhile, Pilk's Battalion, 'or what remained of it', reached Singapore. The lucky ones were sent to a camp to rest, the less fortunate were re-acquainted in hospital.

'Colonel Lywood was admitted suffering from malaria. He

came to my ward twice to see me but he was undoubtedly a very sick man. Major Cubitt was also admitted with a poisoned hand.'

On 31 January the *Singapore Free Press* ran the headline: **'Fighting Now 20 Miles From Johor Bahru'**. The Raffles reprised their ad, while the Seaview spruiked their 'popular concert by Rellers' Band', a busy bunch who also backed up at the Adelphi on Sunday afternoon.

That same day, the balance of Westforce cleared Johor Bahru, withdrawing the last troops over the Causeway: the Australian 22 Brigade and the Gordons were followed by the Argylls, whose two surviving pipers filled the morning air with *Blue Bonnets over the Border*. Engineers then blew a 70-foot gap in the Causeway, rupturing the islands' main water supply. The Battle for Malaya was over. 'On average our troops fought two battles, repaired four or five bridges, and advanced 20 kilometres every day,' wrote Colonel Masanobu Tsuji, Chief of Operations and Planning Staff, 25 Japanese Army.

Yamashita and his three ace divisions were now poised in Johor Bahru ahead of the prize fight — the Battle for Singapore. Building like an angry tropical cloud head were the Imperial Guards, plus 5 and 18 Divisions, the latter two jungle-trained and combat hardened, but fresh. Twenty-eight thousand five hundred troops. Tanks bristled behind them, while 200 aircraft were ready to unleash their deadly loads.

Numerically, the Allies and local troops dug into Singapore numbered around 130, 000. But perhaps the most portentous omen was that trusting Chinese shopkeepers, who'd always allowed colonials in Singapore to operate on a chit system, withdrew their credit.

'In the hospital, blackout restrictions were rigidly enforced. Theoretically all doors of wards had to be kept closed, but the sisters refused to allow this owing to the heat. My door across which my bed stood was never closed except when I was having a dressing renewed.

A Post Office Telegram received by Phyllis on 3 February read 'IMPORTANT. REGRET TO INFORM YOU OF REPORT RECEIVED THAT CAPTAIN HP PILKINGTON WAS WOUNDED

ON 24th JANUARY 1942. LETTER FOLLOWS. UNDER SEC OF STATE FOR WAR.'

'After the first ten days, life began to be more rosy. My temperature dropped to a steady 101/102, food again interested me. Monty Smyth visited me twice between January 29 and February 4 and sent off a cable for me, but instead of sending it from the Cable Office he sent it from the local PO with the result that it was delayed and never got through. Another cable was sent by the Padre for me. That also never got through.'

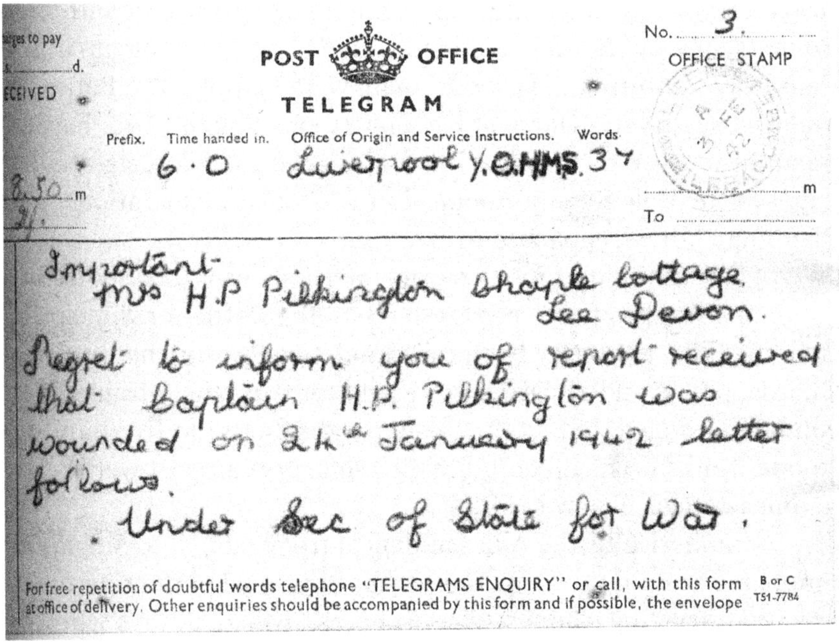

...ges to pay
...d.
...ECEIVED
8.50 m
2/.

POST OFFICE
TELEGRAM

No. 3.
OFFICE STAMP

Prefix. Time handed in. Office of Origin and Service Instructions. Words
6·0 Liverpool Y.O.HMS. 34

To

Important - Mrs H.P Pilkington Staple Cottage Lee Devon.
Regret to inform you of report received that Captain H.P. Pilkington was wounded on 24th January 1942 letter follows.
Under Sec of State for War.

For free repetition of doubtful words telephone "TELEGRAMS ENQUIRY" or call, with this form at office of delivery. Other enquiries should be accompanied by this form and if possible, the envelope B or C T51-7784

Shiploads of troop reinforcements continued to arrive in Singapore Harbour and the straggling *Empress of Asia* was sunk by dive-bombers in the Straits on 5 February. Long-range Japanese guns bombarded Singapore. The hospital was first bombed about 5 February, no one being hurt. Next morning at dawn it was bombed again with a few minor casualties.

The next day saw all civilians evacuated from the northern shores of Singapore, the Naval Base demolished and oil dumps set alight.

On 8 February, the Japanese assault on Singapore

commenced with hundreds of landing craft fording the Straits of Johor from 7 pm. The onslaught was on the mangrove-ridden north-west coast, Percival having deployed his fresh 18 Division to the north-east in response to the classic feint via Pulau Ubin. Barely a strand of barbed wire protected Singapore's 120-kilometre coastline as the Japanese headed for the positions where Gordon Bennett's Australian 8 Division — many of whom he described as 'recruited on Friday, sailed on Monday' — was dug in. Some reinforcements had literally never fired a weapon of any sort.

The first wave of small boats contained 4000 Japanese and at 8 pm the 2/4 Australian machine-gunners let rip. At first there was success, with 40 percent of Japanese casualties (of a total of 1713 killed and 3378 injured) on Singapore falling at the landings. Wave after wave eventually overwhelmed the resistance. The north-west was breached, the Japanese making a break down the western side toward Buona Vista and Kent Ridge.

The Alex was by now impossibly crowded. Sergeant Edward Lunt, RAMC, the hospital steward, put the number of patients at 900, with 180 staff. The overflow was sent to makeshift hospitals in the Goodwood Park Hotel, Victoria Theatre, Cathay Building, even St Andrew's Cathedral. The precaution of laying a large red cross on the hospital lawn had already been taken, but a Bofors gun near the sisters' mess surely made it a legitimate target.

'On February 8 two sticks of nine bombs each were dropped across the building. The sisters' quarters were demolished while several slight casualties were caused in the mental wards. I was operated on again.

It should be remembered that the vicinity of the hospital was plastered with Ack-Ack guns of all calibres which would fire on the slightest provocation and, as close by was the Alexander Ordnance Depot, Alexander Barracks and the oil tanks, these made a target which was being bombed almost continuously. The siren was only three hundred yards from the hospital and we were all thankful when it was put out of action by a direct hit.

On February 8, for the first time, all the big guns of Singapore and Blakang Mati opened up at night on the enemy positions. The noise of the guns and the explosions of their shells, many of them fifteen inch, was almost unbearable, while the concussion on my eardrums drove me nearly frantic. The sisters and ear specialist did all they could with bandages and cotton wool but nothing was really effective.'

That night the matron, Miss Jones, asked for eight volunteers among the nurses to remain in the hospital should it come under occupation by the Japanese. She started by placing her name at the top. Sisters walking to work at the hospital could see Japanese snipers in the trees. 'They made no attempt to injure us, maybe showing respect for our uniform,' said missionary nurse Edith Stevenson in *The Last Lap*.

'By February 9 the Japs had driven back the front line to within a mile of the hospital, when much to our sorrow all the whisky and beer was destroyed to prevent it falling into enemy hands and as a precaution against drunken Japs molesting our sisters. From now onwards the area round the hospital became a battle ground with constant rifle and mortar fire.

On Tuesday 10 fifty of the sisters, much against their wills, were evacuated. I also heard that Campbell-Orde had been killed that morning in the bombing of Selarang Hospital. On Tuesday evening we had our last slap-up meal as, by Wednesday morning, the hospital had been completely deserted by its Chinese and Eurasian staff who had bolted to a man. Our meals now consisted of hunks of bread, butter and bully with mugs of tea. If you couldn't eat it you went without, as nothing else was available.'

11 February marks National Foundation Day in Japan, 1942 notching up 2600 years since the coronation of Emperor Jammu. The grovelling Yamashita had desperately wanted to present Singapore to Emperor Hirohito as a gift on this day. The high ground of Bukit Timah (meaning 'Tin Hill', but militarily known as 'Hill 200') had been taken; the road to the city was open. But, knowing that his own supplies were running low, Yamashita called on Percival to 'give up this meaningless and desperate resistance'. Many Japanese troops were down to their last 100 rounds of ammunition. Percival, reeling, abandoned his Sime Road HQ and bunkered down in Fort Canning.

Brigadier Stringer, commander of medical services for Malaya Force, also pressed Percival to surrender. He argued from a medical standpoint: 9000 on the island were in the care of military medicos, and the outbreak of a hygiene-related epidemic was likely, perhaps imminent.

'On Wednesday 11, twenty-five more sisters were evacuated. The hospital was now under Jap shell-fire and tiles and roofing were flying all over the place. Feeling had come back to my fingers after about three weeks from the date I was wounded. The sisters were overjoyed as it had been the general opinion that I would never use my arm again and might have to have it amputated. I owe my arm to a magnificent bit of surgery at the casualty clearing station by a Captain Smyth.'

On Thursday 12 the battle line was only 400 yards from the hospital.

Percival had ordered withdrawal of his forces into a final protective perimeter measuring just 28 kilometres. Within it were Kallang airfield, Thomson Village, and the vital MacRitchie Reservoir (the Allies' and civilians last source of fresh water). From Adam Road in the east, the perimeter followed Farrer Road, Holland Road, then Buona Vista Road to the west coast. This had obviously been breached, given that the Alex is a kilometre or so south of Buona Vista Road.

'I was operated on in the morning having my plaster reinforced. When I came out of the theatre I found that the rest of the sisters had been evacuated. By midday all patients who were fit to walk had been told to find their way to Singapore. At two o'clock shells were hitting the roof in the vicinity of the officers' wards regularly. All fracture cases were then ordered to be moved down to the centre floor where the NAAFI canteen was situated. Fortunately the lifts were still working so we were all down, some 20 of us, by 1 pm, with 8" concrete walls protecting us. That afternoon the water main was hit. I don't think any of us would care to endure that thirsty period again.'

Elsewhere in Singapore, thirsts were being liberally slaked. Midnight was the deadline given for the destruction of all alcohol, due to stories from Hong Kong on the raping and rampaging Japanese. Raffles and the Singapore Cricket Club partied like there was no tomorrow: for the first time, perhaps, there wasn't. Two hundred cases of whisky were consumed with gusto, many members still signing chits in good faith for that which they consumed. What was not consumed was smashed on the floor under supervision of Capt William Frankland, RAMC.

'Thursday 12 night was passed in an inferno of noise and burning oil wells [tanks]. Friday morning showed that the electric power lines had been hit so that now, added to our thirst, was the intense heat of a room full of feverish patients, with no fans working and no doors open. The RAMC orderlies were magnificent and never ceased carrying out their normal routine.'

Straits Times, Friday 13 February: **'JAPANESE SUFFER HUGE CASUALTIES IN SINGAPORE'.**

That 'Black Friday' Percival argued that honour demanded the fight go on. Wavell, having declared that officers must die alongside their men, climbed on a plane out of Singapore. The Japanese 18 Division encountered the Malay Regiment, Pilk's pre-war mob, around Bukit Chandu (Opium Hill), within eyeshot of the Alex. All their promise as a unit was fulfilled in a devastatingly fierce resistance. The Japanese, although victorious, copped a hiding from the regiment that fought, per their motto, to the very last man.

Midday: the matron and all remaining sisters were advised by the Commanding Officer to evacuate immediately, following a cable from Hong Kong. Reports had it that medical staff in the Salesian Mission on Chai Wan Road were massacred and, on the morning before the surrender, Japanese soldiers had entered the British field hospital at St Stephen's College, torturing and killing over 60 medical staff and injured soldiers.

Hand-to-hand combat raged just outside the Alex Hospital compound. Major Cubitt, his poisoned hand healed, was discharged. Downtown, eyewitnesses reported it was high-tea and haircuts as usual. 'Friday 13 night,' wrote Pilk, 'was a repetition of Thursday.'

Straits Times, Saturday 14 February: **'CITY'S NEW DEFENCE LINE.' 'All evacuation suspended as from today' ('This applies also to women and children').**

The Alexandra Military Hospital lay, unfortunately, in the path of the avenging Japanese and the retreating 44 Indian Brigade.

Around 1 pm the first Japanese troops advanced from Ayer Rajah Road to the north. Dressed in the green uniforms of the Imperial Guards (according to several witnesses), they followed a flag bearer in single file. A soldier, watched by Captain J.E. Bartlett, RAMC, entered the gap of an overlapping blast wall. Bartlett strode out to meet him, arms raised, then indicated the red crosses on his brassard. CRACK! He was shot at point-blank range and fell into the adjacent ward duty room. BOOM! A hand grenade exploded. A company of Guards (if indeed these soldiers were) advanced in short zigzagging bursts.

CO Lieutenant Colonel Craven immediately suggested surrender with Maj Henderson, the registrar, the hospital chaplain, and Major Bull, who were gathered in his office. Bull spotted a sniper through the window, about 200 yards away. He held a Red Cross flag up to the window. The sniper's response zinged off the wall behind him.

Three platoons hit the hospital. From the west, the first group emerged from the railway line behind the nursing quarters. A second came in from the rear at the north. The third infiltrated via the open balustrades to the west of the main reception area near the operating theatres.

Blasts. Bayonets. Blood. Bodies. Bedlam.

Lieutenant W.F.J. Weston, RAMC, emerged from the reception area with a white sheet in surrender. He was immediately bayoneted. Captain Tom Smiley, a keen rugby-playing doctor from Northern Ireland, moved to wheel a patient from the corridor to the theatre. Bullets fired down the corridor put a stop to that. His motioning at the Red Cross insignia only earned him more shots, which injured Private Lewis behind him.

The Japanese then stormed into his operating theatre. Dr Allardyce had already administered anaesthetic to the patient, 28-year-old Corporal Robert 'Bob' Veitch of the Federated Malay States Volunteers armoured car division. Dr Parkinson was bayoneted in the operating theatre but, still conscious, managed to run out. He collapsed in the corridor towards the main entrance. As the Japanese ran out they finished him off with a further two stabs to his stomach. The already-injured Private Lewis took a fatal bayonet jab, as did McEwan and Dr Rogers. The out-of-it patient Veitch – receiving surgery for shrapnel in the ankle -- was mercilessly gored on the table where he lay.

Captain Smiley raised a defiant hand. The bayonet sliced the webbing of his left fingers, deflecting to his groin. The next jab to his heart struck something solid — a silver cigarette case in his top pocket. He fell down, dragging Private Sutton down with him, both playing dead. Satisfied with their carnage, the soldiers ran off. Sutton dressed Smiley's wounds.

An untold number of soldiers had been murdered, mostly in the beds where they lay, unable either to fight or to escape. 'Why were they like lunatics, their eyes, just like lunatics,' said Dr Benjamin Chew, having watched them smash X-Ray equipment.

Craven made his way downstairs, relieved to see Smiley and Sutton alive. They were soon rounded up by a sentry, but inexplicably exempted from the pathetic group of up to 400 who were bound and hauled out from the north wing of the hospital to buildings over the railway lines. Here, they were crammed into three rooms, each measuring no more than three by four metres. Doors were barricaded, windows nailed shut.

The British forces closest to the hospital were the 36 Company Royal Engineers. In the consummate confusion they didn't respond, fearful that their intervention might legitimise the attack.

'It has never been established who the officer was who gave the order to withdraw under cover of the hospital, but whoever he was he broke the first and greatest rule of the Geneva Convention. No blame in my opinion can be attached to the Japanese for the deaths of medical staff and patients during this period of the battle.

For 24 hours we had no food or water. One wretched officer had been given a bedpan at 2:30 and it was not until 12 o'clock next morning that he got rid of it. A sweepstake was run on this like the sweepstakes on a boat!'

That same sickening Saturday, Churchill revoked his 'No question of surrender to be entertained until after protracted fighting among the ruins of Singapore city' edict, and gave Wavell permission to surrender 'when no further result can be gained'. Percival was reluctant.

The *Vyner Brooke*, carrying 65 Australian nurses, was bombed in Banka Strait. Pilk was not to hear the news for a couple of weeks yet:

'But there was our pride when we heard that two of our

sisters had continued helping the wounded when their boat was sinking in shallow water and the Jap planes machined-gunned the survivors mercilessly. And of their heroic journey through the jungle with the survivors.

The matron, Miss Jones, was killed; some we know got safely through; others were bombed at sea and drowned or machine- gunned on the beaches. May God preserve them, they were a grand crowd.'

The *Sunday Times* of 15 February, now reduced to a miniature newssheet, ran the following headlines: **'STRONG JAP PRESSURE,' 'Defence stubbornly maintained. Volunteers in action,' 'The enemy are being held.'**

Meanwhile back over the railway line, many had not survived the cramped, dehydrated, airless conditions of overnight internment. Of those who had, 100 were taken, two-by-two, on a 'water march'. But from behind the latrine block came their fateful cries, smirking guards wiping bloody bayonets. BOOM! The block was shelled — suddenly freed captives sprinted in every direction. Machine guns cut most down. Private Bert Gurd, Corporal George Johnson, Private Sydney Hoskins, all RAMC, Captain Dick Waller, RA, and Sergeant Norman Bryer, RAF, managed to scurry into *lalang*, brush and drainage ditches, and spirit themselves away. They were the only five from that group of around 200 who lived to bear witness to those events.

'At 12 o'clock Sunday the orderlies and doctors still surviving were allowed to return. In a few hours they worked wonders and by 5 o'clock we had all had a meal of sorts, were cleaned up and comfortable.'

At that same time, Percival was being pummelled by Yamashita at the Ford Factory on Bukit Timah Road. Within an hour, their meeting was over, the surrender signed. Approximately 7000 civilians died in the streets of Singapore in that last week of battle. And 8000 British imperial soldiers lost their lives vainly defending it. Governor Sir Shenton Thomas looked out from his balcony and said: 'It doesn't matter about us. It's the people I'm sorry for. It's *their* country and somehow we've let them down.' Some Allied soldiers, unable to face the thought of surrendering, drowned themselves in the vats of the Tiger Beer brewery instead.

'The letters I wrote and sent by sisters from Alexandra Base Hospital never had a chance of getting through. No telegrams or letters reached me before the fall of Singapore.

Again on Sunday night all doctors and orderlies were removed but, this time, we were prepared and things were much better. Before they left the doctors gave the delirious a heavy shot of Morphia. Those who had died had been wheeled, beds and all, on to the verandahs. There was calm and peace again with only the lurid glow of the oil wells to disturb us. Looking back on those three days I am still surprised how cheerful we all kept. There were more jokes and laughs then than at any other period.

On Monday morning our peace was rudely shattered by hordes of Jap soldiers starting to loot the hospital. What they didn't want they smashed. Whatever they wanted they took at the point of the bayonet, even tipping people out of bed and leaving them to lie on the floor. By 11 am the place was a shambles with shouting drunken Japs. I was not molested, being treated more as a curious animal in the zoo, as I don't suppose they had ever seen anyone encased in plaster before. One Jap came up to me, leant his rifle and bayonet against my wounded arm, seized my tin of tobacco, emptied the tobacco out, pulled a canary out of his pocket, stuffed it into the tin, stuffed a lighted cigarette into my mouth and left grinning. Others were not so fortunate and did not get off so lightly. Fortunately I had dropped my gold ring down my neck into my plaster from where it was eventually retrieved by an orderly with a buttonhook.

At eleven o'clock Japanese staff officers arrived and started to beat up their own looters. Pandemonium had raged before but it was nothing to what went on for the next half an hour.

Eventually a grinning staff officer appeared and told us not to worry. He had settled everything and there would be no more looting. He was as good as his word and, in order to enforce his orders, stationed a machine-gun company round the hospital with orders to shoot anyone attempting to loot. These orders were carried out and several looters were shot. This officer very kindly brought back all my tobacco which had been removed in my haversack. Four pounds of it plus quite a lot of someone else's.

From now onwards things rapidly returned to normal. We

got a scratch meal and half a cup of water. Wounds were dressed for the first time for days and I, much to everyone else's annoyance, was given clean sheets.

There were, however, sad and empty gaps in the ranks of the orderlies. The two orderlies that had looked after me so kindly were dead, as was Colonel Lywood, my CO who was one of the walking wounded killed on Friday [*sic*].

On Tuesday 17 the Japanese Divisional Commander [Lieutenant General Mutaguchi Renya, 18 Division] came to the hospital in white gloves to show that he apologised for the crimes committed.

He saluted every dead body where it lay still in its bed and gave permission for them to be buried. He told us that we were now prisoners of the Japanese Emperor and that we should be well looked after, that water should be brought to us and electric light and food. Needless to say none of these things ever happened! What the total casualty roll in the hospital was will never be known but was probably in the region of five or six hundred.'

A more common figure of approximately 250 seems to have been agreed on.

Singaporeans awoke to a new masthead on Monday 16 February, 1942: The *Shonan Times*. The name was derived from the Japanese name for Singapore, *Syananto*, meaning light of the south. The editorial gushed that 'The fortress of the Lion City has fallen like a thunder-smitten oak. The gateway to the unlimited treasures of the East has been forever slammed on the British.' This probably said more about their actual designs than the flimsy 'We're-All-Asians' co- prosperity veneer.

On the island 53,200 British and Australian soldiers, plus around 37, 000 local volunteer regiments and Indian Army officers were now prisoners of war, and finding their way by foot to the Changi peninsula. As POWs, they now officially considered members of the Imperial Japanese Army. Having surrendered, though, they were ranked the lowest of the low.

At Changi Beach, the first Chinese victims of *sook ching* (purity by elimination) were being machine-gunned. Within three weeks, 30, 000 suspects would be executed by the Japanese for being pro-Chinese, pro-British, or suspected members of Chinese secret societies simply having tattoos.

'By Thursday 19 we were getting reasonable meals of biscuits and tinned food with two pints of strong sweet tea and two pints of water.'

On 21 February War Office Casualty Branch wrote to Mrs H. Pilkington: 'Madam, With reference to your letter dated 12 February, I am directed to inform you, with regret, that no information is at present available regarding your husband other than that conveyed to you in War Office letter dated 10 February 1942, nor is the department in possession of any information as to whether your husband was evacuated from Singapore. Every effort is being made to obtain further information regarding him and you will be notified at once of any news that may be received. You will, doubtless, appreciate, however, that in view of the recent events at Singapore it may be some time before further information can be obtained. In regard to your enquiry about time taken in the transmission of cables to and from Singapore in January last, I am to suggest that you approach the Head post office at Ilfracombe or the nearest Cable and Wireless office for information. I am, madam, your obedient servant. [Signed] Dawkins.'

'On Saturday 21 I was operated on for the last time. This time a plaster jacket manufactured by six doctors near 3⁄4" thick was fitted on me and from now on I was gradually able to sit up. I was taken back to the top floor where I was excellently nursed by other officers who could by now walk about.

On Sunday 22, the Japs ordered the evacuation of the hospital to Changi Barracks and the exodus started on Monday with all the fit people that had by now congregated under the protection of the hospital.

Tuesday and Wednesday everyone who could walk was sent off to trudge the 16 miles to Changi. Some of them had only just had arms amputated, but the Japs were adamant: no transport for anyone who could walk. Many of these were two or three days on the journey. I came across Hughes of Dunlop Estates who had lost his leg. I heard his voice as I passed by his bed on my stretcher say, "Hello Pilk. So you have returned to Malaya after all". To which I replied, "Yes, but I have at least had four years at home".

On Thursday 26 I was evacuated by ambulance, one of the

last to leave the hospital, and with me I took my rubber mattress, pillows, sheets, books, etc., all of which I retained throughout the rest of my hospital life and – with the exception of the rubber mattress – I was still using the remains of the blankets and sheets on my last night in Changi Gaol.'

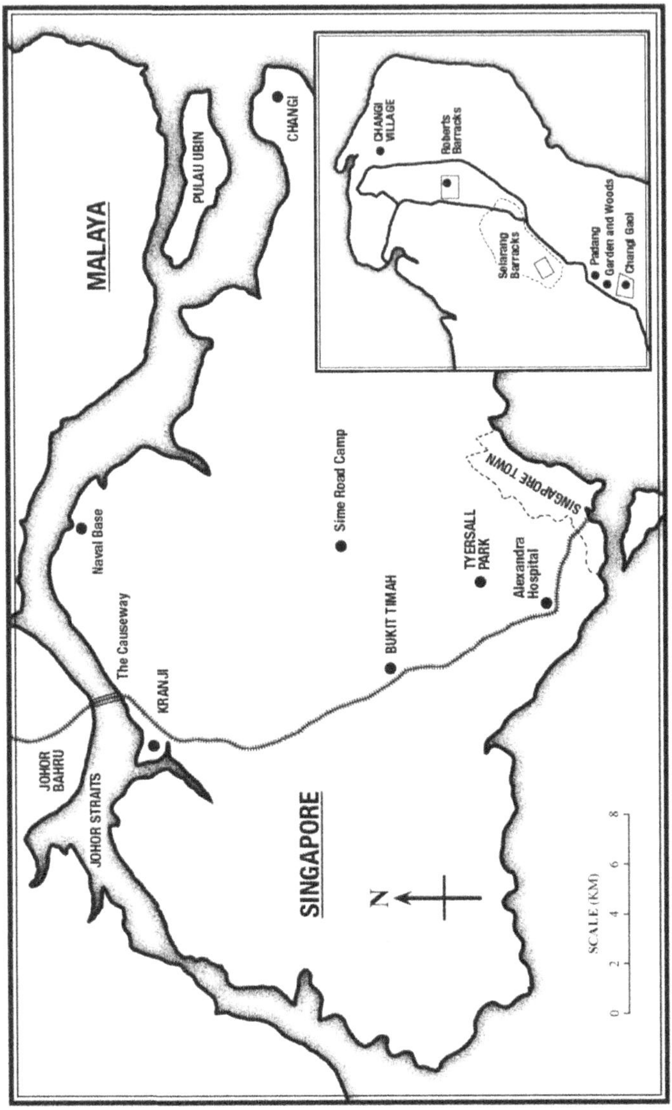
MALAYA
PULAU UBIN
CHANGI
CHANGI VILLAGE
Roberts Barracks
Selarang Barracks
Padang
Garden and Woods
Changi Gaol
Naval Base
Sime Road Camp
SINGAPORE TOWN
TYERSALL PARK
Alexandra Hospital
BUKIT TIMAH
The Causeway
KRANJI
JOHOR BAHRU
JOHOR STRAITS
SINGAPORE
N
SCALE (KM)
0 2 4 6 8

The Alexandra Hospital Singapore, scene of the heinous massacres Pilk lived through on 14-15 February 1942. Up to 300 perished.

The Selarang Barracks Incident, 1942, in which the POWs stood up to their new masters. (AWM 106531)

5 FLIES AND FILTH

Roberts Hospital, Changi, 26 February–10 May 1942

'On my arrival at Roberts Hospital I was carried up three flights of steep stairs to a bed on the balcony of the top floor. This floor, built to accommodate 40 men, had now to hold 200, so you can imagine the squash. Cooking facilities were poor.'

Designed originally for 8000 members of the Royal Artillery, Roberts Barracks now housed many times that number.

Roberts Hospital, so called, was not in actual practice a hospital at all. In peacetime it was the barracks of the Garrison Artillery manning the 15" coastal batteries. It consisted of eight huge concrete buildings which at one time had over 4000 patients in them but had been built to accommodate only some 1000 men. There were beds everywhere, four rows down the middle of the wards, beds on the balconies, beds on the staircases. The overcrowding was appalling but the Japs said we had just to make the best of it.

'Our only meal the first day was one army biscuit and two spoonfuls of weak stew. Breakfast next day 1/5 of a tinned pilchard and biscuit and ½ pint of tea. Lunch 1/30 of a tin of bully and biscuit, ½ pint of water. Tea nil. Supper a saucer of weak stew and no biscuit.

This menu was continued for about a week till finally all our brigadiers said that things had got to improve as the food was in the camp. Tinned stores were then commandeered from units and our food definitely improved slightly in bulk, if not in quality.'

The whole of the 54, 000 British and Australian troops that had capitulated were by now concentrated in the Changi area, those that were lucky in buildings, others in tents and *atap* sheds and some still out in the open.

Fortunately they had been allowed to bring nearly all their personal gear with them, together with huge stocks of tinned foods, but sanitation was inadequate.

'The sewerage pumps had broken down and many men had already developed dysentery.'

Dysentery was nicknamed the 'three day shits', given that sufferers were often dead within that time. In the period from March 1942 to September 1943, 8000 cases of dysentery were dealt with, and 40, 000 microscopic examinations carried out in the Roberts Barrack laboratory. Some internees couldn't handle the misery and ignominy, suiciding by throwing themselves into the overflowing Otway pits: the fetid latrine trenches.

Fly breeding was already in full swing and before long it was to reach shocking proportions.

'During the whole of this period the fly menace had grown acutely worse. Lying on my back on my bed with a home-made swat I would wait my opportunity and slay them in figures reaching up to the hundred mark at one time. The record for our ward being 178 as far as I can remember. A thousand an hour was quite normal going! Mosquitoes got bad and, with no nets, sleep at night was always broken.

At home Phyllis was fretting awfully. On 10 March she placed a notice in *The Times*'Missing' column: 'Pilkington — Captain H.P. Pilkington, Royal Norfolk Regiment, last reported wounded in Malaya, Jan 24 1942. Any information gratefully received by his wife at Woolacombe, N. Devon.' A further half a dozen notices relating to Malaya, Bangkok and Singapore were placed by distraught wives and parents.

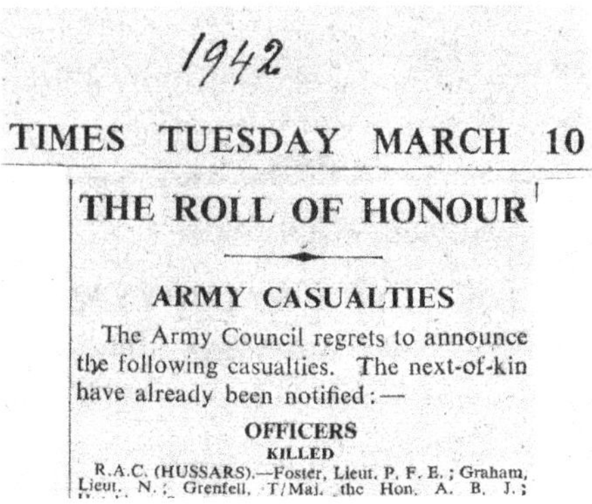

1942

TIMES TUESDAY MARCH 10

THE ROLL OF HONOUR

ARMY CASUALTIES

The Army Council regrets to announce the following casualties. The next-of-kin have already been notified:—

OFFICERS

KILLED

R.A.C. (HUSSARS).—Foster, Lieut. P. F. E.; Graham, Lieut. N.; Grenfell, T/Maj. the Hon. A. B. J.;

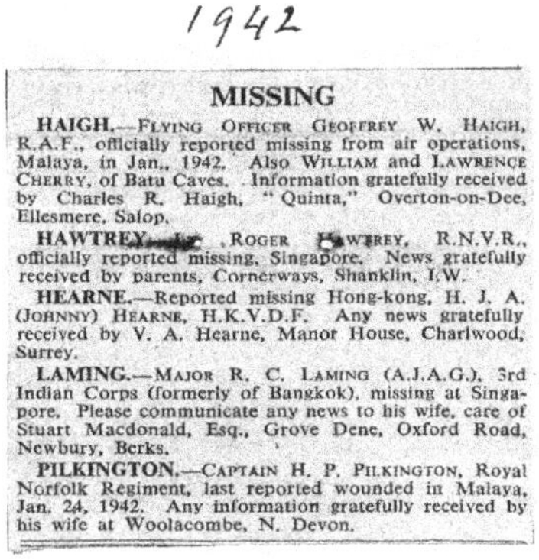

1942

MISSING

HAIGH.—Flying Officer Geoffrey W. Haigh, R.A.F., officially reported missing from air operations, Malaya, in Jan., 1942. Also William and Lawrence Cherry, of Batu Caves. Information gratefully received by Charles R. Haigh, "Quinta," Overton-on-Dee, Ellesmere, Salop.

HAWTREY.—[illegible], Roger Hawtrey, R.N.V.R., officially reported missing, Singapore. News gratefully received by parents, Cornerways, Shanklin, I.W.

HEARNE.—Reported missing Hong-kong, H. J. A. (Johnny) Hearne, H.K.V.D.F. Any news gratefully received by V. A. Hearne, Manor House, Charlwood, Surrey.

LAMING.—Major R. C. Laming (A.J.A.G.), 3rd Indian Corps (formerly of Bangkok), missing at Singapore. Please communicate any news to his wife, care of Stuart Macdonald, Esq., Grove Dene, Oxford Road, Newbury, Berks.

PILKINGTON.—Captain H. P. Pilkington, Royal Norfolk Regiment, last reported wounded in Malaya, Jan. 24, 1942. Any information gratefully received by his wife at Woolacombe, N. Devon.

Other than the odd POW work parties, Westerners had all but disappeared from the landscape in Singapore. All European civilians – 2585 of them including 354 women and 49 children – were interned in Changi Gaol. Everyone, including Governor Shenton Thomas, were made to walk to their designated camps in order to deliberately humiliate them in front of the Asians.

'There was unspoken pride we all shared when the civilian women internees were made to march in the mid-day heat from Singapore to Prison, some 12 miles, which they did, singing *For There Will Always Be an England.* So high was their morale that it even affected a Jap guard who turned on a sniggering onlooker and gave him the beating up he deserved.

During this period the officers in the beds on either side of me died, both of secondary haemorrhage. The padre three beds down became delirious. Not a restful existence.

About 5 March, Colonel Houston, the surgical specialist, opened a special ward for convalescent cases in a small block of offices. I was one of those selected and he took me over from Major Webster who had been so good to me.

I was given a bed in a corner room for five. I had a corner all to myself. By now I could walk with help so got down quite easily. Other officers very kindly brought my rubber mattress and few belongings. In this ward I found a Major Mike Louth, a most charming man who later became my greatest friend and companion. He had been hit in the head by shrapnel which had paralysed his left leg. He was for the whole four years one of the most cheerful people you could meet.'

The rampant Japanese, having landed their troops on Java on 1 March, had 100, 000 Allied troops surrender to them a week later.

'By now the Japs had put us on to rice rations but unfortunately our British cooks had then not the faintest idea how to cook it and refused to be taught. The result was a gluey, unpalatable sour mess. The rice served was often forty-eight hours old and should never have been given to anyone.'

POWs soon learned the beauty of rice: you could simply add *anything* to it and it was largely improved!

'Two days after I had been moved down I developed acute

tropical enteritis and put on so-called slops which consisted of a mug of milky tea three times a day. After about a week I was put back on a rice diet but, as the acute enterical symptoms again broke out, I was put on a special diet which consisted of the following:

Breakfast – biscuits, 1/4 of a sardine, 1 pint tea. Lunch – 2 biscuits, 1 pint tea. Supper – 1 pint weak vegetable stew, 2 biscuits with perhaps a piece of cheese or some jam, 2 spoonfuls of rum (until it ran out).

This was all I got till mid-April when I was put on a very special diet of meat, fish, eggs, etc. for a fortnight to pull me up.

Mike and I used to supplement our meals with whatever we could buy. A small tin of pineapple 4. Tin of jam 4. Cheese 2. Butter 2. Griffiths used to bring me whatever he could lay his hands on. Sergeant Bullen swapped my cigarettes for whatever he could get for me and my men and brother officers brought in all they could. But still it was a period of acute starvation and, what's more, the diet never cured the enteritis. That I had for nine months.

Owing to the shortage of orderlies I arranged for Massingham, an old member of my company, to come and look after me. This he did most faithfully, bathing me, washing my clothing, and doing all the other little odd jobs that wanted doing.'

The Americans had surrendered at Bataan, Philippine Islands, in what was arguably their largest drubbing in the Pacific. Douglas MacArthur, like Gordon Bennett, had scarpered to Australia with the promise: 'I shall return'. Shortly after, Chinese forces had joined British forces in Burma, but by the end of the month the Japanese had cut through Burma to reach Lashio, significantly cutting off land routes to China. On 5 May Japanese troops entered Yunan province via the Burma road.

By now, the Japanese had achieved their immediate war aim of gaining access to raw materials such as tin and rubber to fuel its expansionist ambitions.

'During this period I met many old friends: Maurice Cole, Stan Woodroffe, Brownie, Shotter, Gates, Stevenson and many others, including all the officers of the Malay Regiment.

As soon as the X-rays were working I was X-rayed twice but the plates showed that up to then little or no bone had formed. I was therefore put on Radiostoline, with electric treatment to stimulate the muscles and nerves.

On 9 May it was decided to close down the officers' convalescent surgical ward and transfer us back to the main ward. Feeling much fitter now and able to walk about with ease, I found the stink of suppurating wounds and the smells in the ward more than I could bear. Here Jacko, our MO, stepped in and, backed by Colonel Houston, got permission to remove me to the quarters being occupied by the battalion in Quadrant Road.

He promised to make himself personally responsible for me and very conscientiously he carried it out. On 10 May, escorted by Massingham and six of my men carrying my bedding and kit, I said goodbye to the hospital with no regrets. I passed through the no-man's land between the hospital cage to the 18 Division cage and was honoured by a salute from the Indian sentry who the Japs had put on to guard us. I was still in my plaster jacket.

I grew a beautiful beard which tickled abominably so I was quite pleased when the Brigadier ordered me to shave it off.

Monty Smyth was particularly kind to me, bringing me up a Burma cheroot every evening. Nothing could have given me more pleasure.

Looking back on those weeks at Roberts Hospital one realises that our medical authorities had to cope with great difficulties, but at the same time I shall never feel satisfied that all that could have been done was done and I am convinced that all we patients suffered unduly from petty jealousy and greed amongst the higher officers administering the hospital.

I went to Roberts Hospital with $67 but had to expend $37 of this precious hoard on food to keep me going. With the balance Jacko collected tins of soup, fish etc. on which he built me up over the ensuing months.

The doctors and surgeons were magnificent and for them I have nothing but praise.'

6 A HAPPY INTERLUDE

Quadrant Road, Singapore, 10 May–31 October 1942

'I should explain that Changi POW Camp covered an area some three miles square which was divided into four areas:

1) Changi Village and Barracks in which were concentrated 18 Division.

2) Southern area which was that which had previously been used by the Infantry garrison but now housed all the troops that came directly under Malaya Command, such as Malay Regiment, the Manchesters, Gordons, Leicesters, Surreys, etc.

3) The Gardens and Woods area, previously known before the war as Birdwood Camp, where all the Indian Army British personnel were put.

4) Selarang which was taken over by 8 Australian Division and other Australian Corps troops.'

In round numbers, Australia had 14, 000 men captured on Singapore, and a further 7, 000 in Java, meaning that for every two helpless and hapless Aussie prisoners of the Japanese, there was only one war-ready soldier to defend home soil in Australia.

'Each area was separated by a triple dannert wire fence with gaps on the main roads guarded by Indian sentries. To pass from one area to another you carried a flag with queer Japanese characters on it. These flags would take you anywhere even outside the wire.

All Indian and Jap sentries had to be saluted. Failure to do so resulted in a murderous beating up. All parties had to move about in military formation accompanied by an officer. We had heaps of freedom providing one observed the rules and could walk many miles unmolested. We ourselves were responsible for our own roll calls and had to guarantee that no holes were made in the wire! At the gates we had to place our own sentries who were responsible for seeing that no one without an authorised pass moved from one area to another. No movement was allowed between lights out at 10 o'clock at night and reveille at 6 am. We were allowed our own bugle calls and, by a Japanese order, administered our own discipline, using our own Army Act and King's Regulations. The General was entitled to court martial anyone and authorise terms of imprisonment in our own detention barracks. On the whole the system worked well.

The buildings occupied by the remnants of my battalion, in all some 18 officers and 650 other ranks out of a total of 37 officers and 940 OR [other ranks], were:

1. The NAAFI [Navy, Army and Air Force Institutes] Canteen of the Changi garrison. A large concrete building which was being used as stores, QM office and kitchens and dining hall.

2. A large wooden hut with *atap* roof with normal accommodation for 40 men now housing 80.

3. A shed built by themselves.

4. A row of 10 married quarters. These had only just been completed and were of the most modern brick and concrete type.

Each house contained downstairs a large pantry, two small verandas and a huge sitting-room/dining-room of 40 x 20 foot. Upstairs, a bathroom with long bath and shower, WC, two small verandas, one small bedroom and one large one 30 x 20 foot. At the back was a kitchen, store and one living room, connected to the house by a short covered way. Each house had a small plot of garden in front partitioned off by hibiscus hedges and planted with flowers.

The officers were occupying the last two houses in the row, the big room in one being used as a dining room and in the other as an anteroom. They had made sofas and chairs out of boxes and car seats, tables they had looted as well as dining room chairs, and had very kindly pinched a long chair for me.

To find these quarters spotlessly clean with their little gardens full of flowers, after the stench and squalor of the hospital, was like walking into Paradise. I was given one of the little rooms to share with John Calder and which Jacko had fitted up with an iron bed and good biscuit mattress which he had obtained from God knows where.

Massingham soon made me very comfortable and, for the first few weeks, I would be in my bed only coming down for meals, while in the evening I lay in the long chair till I was sent to bed at 10 pm. We had no lights whatever so had to sit in the dark. It is surprising what fun and arguments we had.

I was now on Jap as against hospital rations. Our meals were as follows:

Breakfast: Small plate of plain rice with teaspoonful of sugar. A small rice rissole flavoured with something, generally stolen coconut.

11 am: 1/2 pint of tea — no milk and no sugar.

Lunch: Small plate of rice, grated coconut and a rissole made with very indifferent curry powder.

Tea: Half pint of tea. Large helping of rice. Meat or vegetable stew. A sweet rice rissole, perhaps with jam. 1 pint tea.

To supplement this diet we later had hibiscus leaf cabbage, tapioca leaves and choice bits of grass. The Japs at first allowed us 4 oz [ounces] of meat per week but later this was cut down to 2 oz To supplement the meat we ate snails in various forms — they became so precious I started a snail farm.'

May 1942 saw the balance of the war shift slightly, signalling a number of developments and a change in the conditions for POWs and civilians. The Battle of Midway in June was a real turning point — the vanquished Japanese Navy could no longer provide aircraft carrier cover on the Indian Ocean side of the Malayan peninsula, meaning merchant shipping to Burma was vulnerable. As 1942 drew on, the Japanese were losing one ship every three days. Somehow, an overland route would have to be forged.

Under the Geneva Convention, which was signed but never ratified by the Japanese, Other Rank POWs were allowed to be compelled to work, so long as the work was not heavy, dangerous, or directly concerned with the enemy's war effort. On 14 May 'A' Force work party comprising 3000 men was assembled and sailed from Singapore to Burma, from where the British were in full retreat to India.

'At the end of May the plaster was cut away from my shoulder leaving it bare to the air, the wound now being healed and I was given a daily dose of electric vibration treatment. At the end of June the whole plaster casing was cut away and my arm put into an aeroplane splint. This was much more comfortable and cooler. In the middle of July this was discarded and I was told I was to work the joints loose. I couldn't even bend a finger, let alone the elbow. Jacko gave me a volume of the *Encyclopaedia Britannica* to hold. With the weight of this book the joints gradually opened up till at the end of about three weeks I could use my hand and forearm easily.

The Brigade gardens started producing vegetables, mostly *kang kong* and sweet potato leaves, but they were a godsend even though one normally only gives them to pigs as a last resort. Occasionally we got a handful of beans or some green vegetables from the Japs.

For the average person used to a European diet, the food was at this stage literally murder by starvation. None of the sick men were able to recover and beri-beri rapidly attacked us all.

Our Brigade was undoubtedly worse off than most as, at the capitulation, they had been made to burn all their money and could not therefore buy over the wire as others did. To counteract this, as we were still getting a cigarette ration of 20 a week, we sold our cigarettes at 2 for ten and built up reserve funds, bought food and gradually made life bearable.

We were also sending large working parties down to Singapore to work on the docks. They did some noble looting and would return with full packs of every imaginable type of tinned food and even fresh bread that was still procurable. This they always shared with us and I am extremely grateful to all those people who kept me supplied with looted tobacco and luxuries.

The Japs allowed canteens to open which were well stocked with tinned soups, cheroots, raisins, coconuts, etc., all at pre-war prices. In many cases much cheaper.

For weeks before any postcard was written, our British camp commander would make endless representations to the Japanese asking that regular communication with England should be arranged for. For several weeks before the postcard was actually issued they would be pointed out to us by the Japs, stacked up in their guard room, usually with the remark: 'You OK. You write postcard.' Finally about midday they would be issued with orders that, if they were not returned by that evening they would miss the boat and that if any postcard did not conform with their regulations, those of whole units would be torn up.

Every postcard, after it was issued, was checked by a British officer before being handed to the Japs. In spite of the great care taken, we heard that the Japs in a fit of temper had found some trivial excuse to burn or destroy hundreds. The postcards having been returned to the Japanese, they would throw them into bags which would remain literally for weeks and months before finally being sent off on a boat.

On 2 June we were allowed to send one postcard home of 24 words. It was better than nothing.' Pilk's postcard appears to have never been received by Phyllis.

On 5 June, at Nong Pladuk, central western Thailand, a small but significant signpost was stuck in the ground. It read '0 kilometre'. Following an agreement with the Thai government, this would be the starting point for the Japanese' strategic railway link to Burma. On 20 June, soon after the Battle of Midway the Japanese Army HQ made a formal decision to proceed with the Thai–Burma railway using their unexpected windfall of POW labour. In all of the Imperial Japanese Army planning for the railway link (commencing in 1939, but mooted as early as 1920) never had they factored in that they'd have over 100, 000 POWs at their disposal.

Surveys completed in 1942 resulted in a calculation that 60, 000 workers (10, 000 of them POWs) would require five years to complete the railway from Nong Pladuk, Siam, to Thanbyuzayat, Burma, a total of 415 kilometres.

The British had proposed a Thai–Burma Railway connection as early as the 1890s and surveyed the route in first decade of the 20th century. However, sovereign sensitivity issues aside — it was not their country and they were dealing with a proud monarchy that had never bowed to imperialists — they'd dismissed it as unfeasible, given the terrain and the tropical diseases endemic in the area.

A point to note before returning to Pilk's memoirs: Thailand was known as Siam for centuries. While the dominant colonial powers of the Great Age of Discovery — the Dutch, Portuguese and French — sought favour with the kingdom, an anti-Western coup in 1688 marked a century and a half of virtual seclusion from Europe. By yielding Malay-dominated territory in the south to the British, and Indo-Chinese territory in the north to France, royalist Siam managed to remain the only country in South-East Asia to avoid colonial rule or control. In 1939, the new military-minded Prime Minister, Phibun, embarked on anti-Western and anti-Chinese campaigns. Reflecting this, the more nationalistic name of Thailand came into being. Hence the favours extended to Japan in exchange for not being occupied by Japan's 15 Army.

'Parties of officers and men started to move off to Siam to start construction on that ill-fated railway. Gradually our numbers decreased to 12 officers and 450 men. I now had a room to myself and even the big rooms only had three officers each.

In July I was asked to take over all brigade and unit gardens. This was not an arduous task for as head gardener I had Sergeant Birkett, Lord Burley's gardener, and three officers to help me. We immediately had all the inside hedges up, dug up all the lawns except that opposite our mess and planted them with papaya trees, sweet potatoes, arrowroot, *kang kong*, maize, pineapples and beans and chillies. By the end of August we had more vegetables than we could eat and by October the papaya trees were giving us abundant fruit. Bananas were slower, but even these became a regular supply. No coconut trees grew in our garden so we either had to steal from our neighbours, the gunners, who had – in our opinion – far too many, or go without. We preferred to steal but owing to the watchfulness of our neighbours and the heavy barbed wire they had wound round their trees, we could only do so on the blackest nights, preferably when a heavy tropical storm prevented them from going out. Nothing loath, our coconut party – clad only in shorts, carrying a long ladder – set out to cover the no-man's land in between to the best of the Gunners' trees, which were by the road within a few feet of their front door. The ladder would be hoisted up and Price, the doctor, as the lightest, went up to gather the fruit and drop it down to us, to be carried back to our mess where we ate our fill. So expert did we become that only a few minutes were necessary for the whole operation. But misfortunes did sometimes befall and there were times when the doctor might be marooned in his tree for a considerable time with the ladder and fruit carriers lying flat in the concrete storm drain below. Price was only just rescued in time, for the tree up which he had been left was found to be full of red ants which had discovered his hiding place, while down below inquisitive Gunner officers searched the neighbourhood.

Every Sunday I sent a party out with a handcart to forage for plants, etc. in the neighbouring abandoned kampongs and whole banana and papaya trees were transplanted fruit and all. Later I extended the gardens to take in every available space, both roads, our buildings and the men's so that by the end of September they were never hungry. I also started from one brinjal a brinjal orchard that was one of the show sights of the camp and was eventually extended to all other gardens by my seedlings. Cucumbers and tomatoes we obtained in small quantities, but lack of manure prevented successful cultivation.

At the end of June we pooled our funds and bought eight laying hens and a cock, complete with hen houses and runs for $40. From then onwards we always had at least two eggs a week, feeding the birds on surplus rice from the swill bins. In July we bought 12 chicks which all grew up into good birds. We also bought 20 ducklings at 10 cents out of which I, with Carlton's help, successfully reared 15.'

Come July, the House of Commons gave Churchill and his war effort a 475-25 vote of confidence, as B Force sailed from Singapore to North Borneo. On 9 July Phyllis received a letter from the Casualty Branch, War Office:

'I am directed to inform you that no further information has been received concerning your husband, Captain H.P. Pilkington, The Royal Norfolk Regiment, who according to information available in this Office, was in Malaya during the hostilities which terminated in the capitulation of Singapore on the 15th February, 1942. All possible enquiries are being made to obtain information concerning him. It is hoped that he is safe and, although he may be a Prisoner of War, it is regretted that it is necessary to post him as 'Missing' pending receipt of some definite information. I am, Madam, your obedient servant.'

On 14 July Indian National Congress demanded the departure of the British, who agreed on passive resistance until India's independence. Just over a week later the Japanese commenced their invasion of east New Guinea, leading to the infamous events on the Kokoda Track. For Pilk and his men:

'In July they issued what they called working pay $1/8 per week for officers and $1 for men. It was amazing what a lot one could do with this, plus the money obtained by selling cigarettes.'

Around this time, all Allied POW officers above the rank of colonel were transferred to Japan and Karenko camp, Formosa. Civilian dignitaries, such as Governor Shenton Thomas, were similarly moved.

'When all the generals went to Japan, Colonel Holmes of the Manchester Regiment took command. The Brigadier, as a parting gift, gave us a hurricane petrol lamp, 10 gallons of petrol and six spare mantles and burners. Our dark nights were over and we were able to play cards, the favourite game being called Stooge which was a combination of Hearts and Slippery Sam.

In August, Bidulph of the Chinese Protectorate, succeeded in procuring for me a muscovy duck and drake which I called Olga and Stalin. They were the greatest fun. She laid on an average 10 to 12 eggs a month. Stalin grew to be a large and fierce bird with other people although with me he was as tame as anything. They had a huge bath to swim in, a nice cosy house with glass windows, and a big run. They flew all over the place but always returned.

At this time I had hundreds of doves feeding on the lawn, a lovely sight. Also a lot of minahs became quite tame and would sit on my windowsill chatting away quite happily. Our animals certainly lived a happy and joyous existence. In the evenings we would try to arrange cockfights with a neighbouring hennery. Basil, our bird, soon became champion.

In August a Red Cross ship came in from South Africa and for once the Japs almost played fair, sending us up about 50 percent of its cargo. The food from this ship, when divided out, gave every man something like 24 tins of bully, 15 tins of meat and vegetables, 4 tins of fruit, 100 cigarettes, 5 lbs of sugar, 1 lb porridge mixture, 4 x 2 lb tins of jam, 2 lbs of dried vegetable soup, marmite, lime juice, etc. We literally lived on the supplies of that ship till the following March.

Furthermore, from the clothing supplies of the Red Cross ship we all got a hat and a pair of boots. What a godsend they were.

Normally when cooking we allowed one tin of bully to 20 people. Everything else was on that scale and we would have two or three Red Cross meals a week. From July to the following April, with canteen supplies in abundance such as dried fish, palm oil, sweet potatoes, dried whitebait, ground nuts, towgay, etc., and the gardens producing wonders – plus the Red Cross supplies – we reached the peak of our living conditions.

At the beginning of August I was ordered to report to the surgical specialist who ordered me massage under the only competent masseur they had. This was agonising as he had to break down all the adhesions in my shoulder muscles and get the main muscles in my back, which had been torn when I was wounded, working again. I was thankful when this period which lasted, alternate days, for about six weeks, was over. I was then put onto daily electrical treatment of an hour a day but, at the end of April, there was no response from the muscles and nerves so this was abandoned.

Unfortunately both my ears started suppurating and gave Jacko a great deal of trouble. He cleaned and anointed them himself daily for nearly six months and finally effected a cure. They were at one time in a shocking condition, oozing blood and mucous. The rupture in my right ear had by now healed but I was pronounced three parts deaf in that ear by the specialist.

During this period a daily news service was given, the news being given out in an abridged form to Brigade HQ and then to Battalion. It was generally forty-eight hours old.

In August Carlton suddenly appeared out of the jungle. He had lived in native huts until betrayed by the Malays to the Japs. He had been cut off by a river when the Battalion retreated. He was in good health and I was delighted to see him as I had been unable to trace anything of him. He immediately took over from Massingham and looked after me again until he died.

This first period of captivity was undoubtedly a happy one as far as it could be.'

By September, the Japanese were battling on a number of fronts. On the 15th, US troops repelled the Japanese at Guadalcanal. A week later, the British attacked the Japanese front in Burma. On 23 September, the Australians attacked the Japanese in New Guinea. Within a few days, the Aussies at Milne Bay had inflicted the first land defeat of Japanese, forcing them to withdraw.

'The 18 Division had opened up large gardens, a piggery and some 20 pigs, a metal workshop, woodworkers' shops, central canteens, an electric power station on which allowed each unit 20 light points, together with all the special services of sanitation, plumbing, lighting, etc. Cooking of rice dishes had advanced beyond all belief, all recipes being circulated between units.

By the end of September we were producing more green vegetables than we or the hospital could cope with and were in a position to supply all private gardens too with all the seedlings and cuttings they required. I also arranged the harvesting of outside coconuts which I distributed on a unit basis.

Central and unit libraries supplied nearly every type of book, while those wishing to do so could join the 18 Division University where lectures were given in all the modern languages, farming and agriculture, history, military history, higher mathematics, architecture and English literature. Most of the courses reached a 5th form standard of a good public school.

Clothing was supplied and repaired by a central tailor's shop. When I came out of hospital all I had were some pyjama trousers, one pair of shorts, one pair of boots, one pair of socks and one handkerchief. Thanks to Colonel Hingstone of 18 Division RASC I was completely fitted out with three pairs of socks and shirts, shorts, trousers, shoes, jacket, rain cape, puttees, etc., from the central reserve pool. What he couldn't supply, my brother officers made up out of their own kit.

The big garrison *padang* was allocated into hockey, soccer and cricket playing fields and all who wished could play what games they liked. All fit men had to do 20 minutes PT from 08:00–08:20 every morning.

Three churches were opened; St George's Church under Padre Cordingly in the native mosque, a RC Church, and a second Church of England under the Chaplain General. Services were held regularly and in some cases were accompanied by organ music. All the church furniture was carved and wrought by POWs.

Theatres and concert parties were started by being small local shows developed into the Divisional New Windmill Theatre, with seating for 400. This theatre was gradually equipped with every modern stage device and lighting effect. Amongst the shows put on were *The Dover Road, White Horse Inn, Gentlemen Only and I Killed the Count*. All magnificent productions that would not have shamed a provincial theatre. Sundays were given over to orchestras and concerts of the classical variety, the best undoubtedly being a first-class rendering of 'The Messiah'.

To keep them happy the men had their own little concerts and theatres as well where they got up to their own slap and tickle. The female impersonators were one of the highlights of their shows, Mackwood being particularly distinguished. The cartoons on the walls of the theatre were by Searle, the *Lilliput* cartoonist.'

Sapper Ronald Searle became famous post-war for his St Trinian's girls. Having started in the Royal Engineers with the Territorials, he joined the Regular Army in September 1939 and was posted to Norfolk. On 13 February 1942, under enemy fire, Searle found a copy of the October issue of *Lilliput* amongst debris in a Singapore street, and saw one of his St Trinian's cartoon in print for the first time.

'Sanitation during this period made huge strides. All latrines were now dug to a minimum depth of 20 feet with airtight and light-tight lids and, at the same time, many ingenious flytraps were constructed. Flies and dysentery disappeared together.

The water supply was improved with adequate shower bath facilities for the men. Kitchens were fly-proofed and generally hygiene was instilled into the men. Anti-malarial surveys and work were taken in hand, and all grass was kept mown like a lawn, with the result that an epidemic of malaria was nipped in the bud and the mosquito question carefully watched.

A rather foolish performance of our higher command was the ordering of drill parades and lessons on unarmed combat. The only result was the wearing out of precious boots, the confiscation of steel helmets and a great tightening up all round by the Jap authorities who were not quite the fools they were taken for. Several senior officers were given various terms of solitary confinement in a Jap prison and peace once more returned to the camp.

During this period I read all Dickens with the exception of *Barnaby Rudge*, a number of Thackerays, Kiplings, Stevensons, etc. with Wells' *Science of Life* which I loved and studied for nine weeks.

At the beginning of October, Daykin and some more of our

men from D Company suddenly appeared from KL Gaol. They were all in a poor way with beri-beri and deficiency diseases, but soon recovered their health after a period of our garden produce. These people had also been given up as lost so it was a great relief to see them.'

Pudu Gaol in Kuala Lumpur was built in 1895 by the British at a cost of $320, 000. The X-shaped prison block is notable for its Islamic architectural flavour, and distinctive Minangkabau wall towers, designed by government architect Arthur Norman, who also designed KL's Selangor Club, St Mary's Cathedral (where Pilkington was married), and the Resident-General's residence, the Carcosa. Allied POWs were interred here before the fall of Singapore, numbers of British and Aussies rising to 446 in early 1942, crammed into cage-like cells originally intended for 50, with neither beds nor eating utensils.

'Last but not least our official cemetery was opened. Designed and laid out by Brigadier Duke and Major Brigall, it will be an everlasting and beautiful memorial to those that died here as POWs. The Lychgate, with its beautiful carving, is alone a work of art. Many of the flowering trees planted under my recommendation, when I last saw them, were growing well and will eventually produce a beautiful shady and restful garden.'

7 SIGNING OUR LIVES AWAY

Selarang Concentration Camp, Singapore, 2–5 September 1942

In the middle of August there had been a new change of Jap generals and at the same time the Japanese Government had announced their intention of making all POWs sign a non-escape form — the penalty for the infringement of which would be death.

Under Japanese military regulation, escape was tantamount to desertion, given that the POWs were considered members of the IJA. 'You will never get off this island' was the Japanese mantra. Trouble was brewing ...

By King's Regulations no prisoner is allowed to sign such a form except under duress. By the Geneva Convention no such threat is permissible against a POW, the onus for the prevention of escaping being with the capturing power and the maximum penalty that should be inflicted on a prisoner escaping is three weeks' solitary confinement.

'On 1 September the Japanese General stated that if we did not sign he would shoot four men.

It was realised that he meant to keep his threat but, as the four men were already in his power and were likely to be executed anyhow, nothing could be done to prevent it. On 2 September we were all ordered at 12 o'clock in the morning to be concentrated by 5 o'clock in the Selarang Barracks, a group of eight buildings built to house 900 men.'

Construction of Selarang Barracks commenced in 1936, with the infantry complex completed in 1938, and was occupied initially by a battalion of the Gordon Highlanders. It was part of the Colonial Government's thrust to protect the eastern approach to the strategically important Johor Straits, resulting in Changi being transformed from a mangrove swamp with little other than a few kampongs and police station at the mouth of the Changi River, into an imposing military base. For POW purposes, the Australians had been designated Selarang as their main holding pen.

'At two o'clock the great trek started. All food had to be carried on trailers, firewood, water carts filled and manhandled there. All livestock, bedding, etc., had to be transported.

The distance was about 2 1/2 miles with three good roads leading up to it. All roads were thronged eight deep. By 5 o'clock everyone was inside the prescribed area. The remains of 53 Brigade were allocated half of the second floor of one of the buildings. The floor had to accommodate 600 men and officers. It was built for 36. Each man was given a space 14" wide and 6' long with gangways 18" wide between rows to allow movement.

Griffiths, Jackson, Sindon and myself shared the little lavatory, Jackson sleeping over the pan and I under the washbasin. Calder in the doorway. As far as accommodation went we were lucky. Griffiths had very kindly brought up my mattress so I slept in comfort. Outside, temporary cookhouses were quickly constructed and a small tent allocated to the officers for a mess where we fed in relays. A hot meal was served to all before 8 o'clock.

The Selarang Barrack square was an unforgettable sight. On every inch of ground were small tents and bivouacs, even the roofs of the buildings were covered with tents and awnings. The centre of the parade ground was turned into a huge latrine with trench after trench being opened, while in one corner the ambulances were drawn up to form a hospital with a small tent as an MI room and operating theatre.

It should be remembered that at 12 o'clock that morning there was lavatory accommodation for 800 men. By midnight the REs had constructed sufficient for 15, 000. A magnificent feat! As soon as we were all in the Japs posted machine guns every 50 yards, generally under the good-natured advice of our men. Indian guards were also posted who quite good-humouredly spent their time chasing chickens that had run across the back road. The first night passed without incident. Everyone very merry. On the morning of 2 September officers commanding the various groups were taken to the beach where they were told that they were required as witnesses to the execution of four POWs. These four men had been recaptured in an exhausted condition and very ill. They had been brought back to Roberts Hospital where, after two months' treatment, they had begun to recover. The night before, the Japs had dragged them out of hospital and taken them to Changi Gaol, where they had been tried by a summary court martial and condemned to death.'

These four were the Australians Corporal Rodney Breavington and Private Victor Gale, and Britons Privates Harold Waters and Eric Fletcher who had escaped by rowing boat from Singapore hoping to be picked up by Allied warships. Alas, they were intercepted by the Japanese a couple of hundred miles out and returned to Singapore to be made an example of after convalescing in Roberts Hospital. The execution party made their way to Changi Beach, overlooking Pulau Ubin.

When they arrived at the beach the Japanese kept them waiting for an hour in front of a volunteer Indian firing squad. Eventually the order to fire was given by Captain Dillon, a renegade Indian officer. The Indians deliberately shot to wound and not to kill. It took something like 20 rounds per man to kill them. The Japanese stood by laughing.

Breavington made desperate entreaties that as leader of the group it was he, and he alone, that should be shot. Once wounded he pleaded to be finished off, saluting all the while.

'This day the water supply which came from two wells began to give out and I, with eleven other officers, each took a two hour spell rationing the water to various units. Eventually the REs fixed a hand pump to the main and we were able to increase the ration from half to one gallon per man. This did not allow for washing but was adequate for drinking and cooking.

The day passed uneventfully. There was no communication with the outside world.'

By this stage all American and British films had been banned in Malaya, replaced by Japanese films which proved popular. The Chinese-born Shaw Brothers were paid a $350 'salary' (in Japanese currency, of course) for the privilege of screening Japanese propaganda movies and some Indian titles. Cathay Cinema in Singapore was operating again, providing much-needed escapism to the population. Horse racing in KL had also been resumed.

'The morning of 4 September over 100 dysentery cases were reported, as well as malarial ones. Requests to the Japs to be allowed to remove them to hospital were refused. In the afternoon a man went down with acute appendicitis. Again permission to remove him to hospital was refused and he was operated on in the MI tent within a few yards of the latrines. He lived to tell the tale.

Further the Japs said that if anyone died they must be buried in the latrines.

By evening further dysentery cases were reported, firewood was running out, water was short and the Japs refused to allow any rations to come in. Still everyone very cheery and the hens laying well.

On the morning of 5 September dysentery and diarrhoea were becoming epidemic so that Colonel Holmes, under the advice of the medical officers, asked for an interview with the Jap Commander, then handed over the forms which we signed:

"I, the undersigned, hereby solemnly swear on my honour that I will not, under any circumstances, attempt to escape."

This was contrary to the code of the British Army, which stated that it was the duty of every serviceman to try to escape. Colonel Holmes, the senior British officer in charge, encouraged men to sign in pencil, so as not to be binding under Allied law. Many others signed with pseudonyms such as Mickey Mouse, Winston Churchill and Ned Kelly.

The final threat of the Japanese to move the 4000 sick and the whole of Roberts Hospital in as well undoubtedly greatly influenced Colonel Holmes. In fact the hospital authorities were all packed and ready for the move. If it had taken place there would have been a mortality of at least 50 percent of the total number incarcerated.

At four o'clock the great trek back home began, and was a repetition of the outward one. I was sent back in the water truck, so tired I could hardly keep awake, but we were thankful to be home again and celebrated that evening with an extra large meal.

The chickens and ducks were also glad to see home again as we had had to keep them in a box the whole of the time.

Some of the official statistics for this incident:

Area of concentration 170 x 240 yds Pop. Over sq. mile 750, 000 Inhabitants 15,301 soldiers

Livestock 9 goats, 342 hens, 111 ducks, 3 monkeys, 1 parrot

Soil excavated for latrines1,500 tons Weight of soil to be dealt with per day 6½ tons Size of latrines 20' x 2 ½' x 16'.

No. required 2 per doz.

A curious sidelight to this incident was that a week later we were told that our six months' period of punishment for taking up arms against the Japanese Emperor was over and that we would now be treated as prisoners of war. Further, we were to be paid on the scale of Japanese officers, with the exception that we should only receive $30 per month instead of $124 we should really be entitled to! The rest would be used to pay for our food, lodging and clothing while any left over would be invested in Japanese War Bonds for us. We were only too thankful to get any money at all and at a meeting of officers it was agreed to hand over $10 of our $30 to a central fund to buy essential foodstuffs for the company. The balance of $20 was ours to do with as we liked so we paid $8 into our Battalion Mess. I gave Carlton $2 and had $10 for myself.

The CMF, as the camp messing fund was called, also had all the profits from the canteen so that, following medical advice, each man received a basic ration of whitebait, rice polishings for beri-beri and palm oil for fats per day. The results were most satisfactory and the general health of the camp greatly improved.'

8 A GARDEN PARTY OF SORTS

Quadrant Road, Singapore, 1 November–20 December 1942

'In the middle of October the Japs ordered a party of 6000 men and 200 officers to proceed up country to build the Bangkok/Rangoon Railway. They left on 27 October.'

This group was known as V Party, under Lieutenant Colonel Stitt, and ended up around Chung Kai in Kanchanaburi, Thailand. (It is interesting to note Pilk's nomenclature pertaining to the railway.)

'From those of 6 Norfolks left I had to send three officers and 120 other ranks out of a total of six officers left and 200 other ranks. Griffiths and Calder went with this party.

On October 30 a reshuffle of accommodation took place, Brigade Headquarters coming to live with us. On 2 November the Japanese ordered a party of 300 officers to proceed up country. All officers' names were put in a hat and drawn for. Mine was one of them but the hospital authorities refused even to consider my going. Major Cubitt went with this party. The result was that by 5 November only a fraction of the officers were left and not many men, and those mostly unfit.

It was decided therefore to group all the officers of the Infantry Brigade in our row of houses. We the 6 Norfolks retained our end two buildings but had attached to us two RAMC officers and one padre. I was also given under my command 12 RAMC orderlies and all the ordnance and RASC personnel of 53 Brigade.

There was little difference during this period in living. The main difference was outside as we now had to supply working parties for the Jap market gardens at Tanah Merah and for the forestry party who were cutting down trees for firewood. No man who was fit could be spared for essential sanitary works at home, and this now had to be carried on by unfit men who did a very good job.

Owing to the shortage of officers it was agreed that all

officers, irrespective of rank, should take their turn in going out with these parties. This meant that I had to go to the forestry party once a week and the garden party twice. The other three days I attended the hospital for treatment.

The garden party consisted of marching 100 men out to the fields at 8 o'clock each morning and staying with them till 5 o'clock when we marched home, a distance of about 2½ miles each way. At the gardens we built fires and cooked our midday meal, thickening up the stew with garden produce stolen for the purpose. Once we were out there, I could do as I liked and spent my time in an advisory capacity on anti-malarial drainage and gardening in general. I thoroughly enjoyed these outings.

The forestry party consisted of taking 50 men out to the working site who felled trees and stumps, were loaded on to trailers made out of old car chassis and dragged home, a distance of about three miles over very rough ground. This work was nearly always completed by 1 o'clock and I enjoyed the outings.

Unfortunately the glare of the sun, combined with the beri-beri I contracted, badly affected my eyes with the result that the oculist forbade me to go out in the sun unnecessarily and I was not allowed to read. These conditions lasted till I went up country to Siam in May and, during the whole of that period, I was excused all duties. My ears were again badly affected and were suppurating badly, but thanks to Jacko's care, the inflammation was always kept in check. Being unable to read I fell back on sketching and studying architecture, many very excellent and lovely books being lent to me.

With the departure of Major Cubitt, our unit now came under the command of a Colonel of the ——— shires [full name omitted by Pilkington], a man I could not get on with as he was pigheaded and always interfering with my administration. At one time he threatened to put me under arrest, so I reported the matter privately to Divisional HQ and heard no more about it; from then onwards I was left in peace.'

The International Red Cross committee received the first list of prisoners from Singapore in November 1942. Captain H.P. Pilkington was listed among them. A headline in the Daily Telegraph, Saturday 7 November was naively optimistic: **'Japan's Uneasy Hold on Malaya. Escaped Briton's belief that the**

peninsula could easily be reconquered.'

'Early in November the Japanese decided to tighten up the roll call parades when we all had to parade at 8 o'clock and 7 o'clock on the main *padang*, the officers under Colonel Ling. These parades were quite senseless and very irksome and, as they were in no way a check on the numbers in camp, quite useless to the Japs. We, however, had an amusing time pulling Ling's leg — when we numbered it was perhaps in the number of ships the Japs had lost in a naval action such as one cruiser, two destroyers, etc., which made Ling livid with rage but he was quite powerless to stop it.'

On 15 November, the Japanese lost 28 warships at Guadalcanal which would have provided rich fodder for numbering.

'When my eyes got bad I was excused these parades altogether which was a relief. I was given charge of the combined cook house for all 53 Brigade and, with Sergeant Melton a cook sergeant, completely rebuilt the kitchens with flagged floors, proper drainage and fly- proofed preparation rooms. We were both extremely proud of the result. The men also were pleased and said that they had never been so well fed before, which was satisfactory. On the whole this was an uncomfortable period for, although I retained my private little room, there were now nearly 100 officers in our block, often squabbling and nearly always making a noise till midnight, making sleep impossible till they had gone to bed. It was during this period that we were attached, unknown to the Japs, to the electric light mains of Singapore! We were allowed four lights which I apportioned two to our ante-room, one to the kitchen and one to the batman's living quarters. Evenings were now very much brighter.

In late November I purchased 12 more small ducklings which attracted the attention of a wild cat that used to steal and kill them during the daytime. Charles Wood constructed for me a most ingenious and simple trap which, baited with a dead duckling, did the trick. The cat when caught proved to be a wild tiger cat, about twice the size of a full-grown Persian, very heavy and very strong. It was in fact a complete replica of a tiger, Blang, for instance. I drowned it in its cage in a *tong* [jar] of water.'

(Blang was a tiger kept as a pet by Pilk's neighbours, the Ryves, in Negri Sembilan. It lounged around the house and enjoyed drives in their open-top car, like any normal household animal, but on at least one occasion emptied the main bar of the Selangor Club in double-quick time when taken in there on a leash!)

The first bag of British mail arrived in Changi on 11 December (the Australians had to wait till March 1943 before 12 bags of 80, 000 letters reached them), to much cheer and clamour. This had necessarily come via a most circuitous route: from the UK to the Red Cross in Geneva, Istanbul, Moscow, then on a trans-Siberian train to Korea, on to Japan, thence by boat to Singapore. Mail to and from Australia essentially did the same thing in reverse, tagging on the extra sea journey from England to Australia.

Four days later, a Post Office Telegram to Mrs. HP Pilkington was received: 'Important hand delivery. Chapel Cottage, Lee, Ilfracombe. Special report received that Captain HP Pilkington of Royal Norfolk Regiment is a Prisoner of War. Letter follows shortly. Under Secretary of State for War.' Delayed as this news was, it was Phyllis' first confirmation that he was at least alive. She placed a notice in the 19 December issue of *The Times*' Personal column: 'Captain Hugh Pilkington, Royal Norfolk Regiment, previously reported wounded and missing in Malaya, now reported Prisoner of War'.

That same day the British advanced 40 miles into Burma in a drive to oust the Japanese from 'their' colony. Elsewhere on the peninsula, things were getting bleaker. Rice was at a premium, rising by year end, up to 12 to 15 times pre-war levels in Singapore and Penang. Japanese occupation currency was forsaken in a revival of bartering. There was not much confidence in their 'banana money', so named for the depiction of that fruit on the front of their notes.

'We were by now getting eggs four times a week as our pullets had started laying and with the reduced number of officers, our turns came round more frequently. Terry Daykin took over sole charge of the hens and made a very excellent job of it. We never lost one hen through disease and Basil, our cock, won all his fights, much to our satisfaction. My muscovies were by now tame and caused everyone the greatest amusement as you can imagine.'

10
10
THE
JAPANESE
GOVERNMENT
PROMISES TO PAY THE BEARER
ON DEMAND
TEN DOLLARS
MP
MP
10
10
10

The Japanese introduced their stamps and money into Malaya 1942-5.

Right: Pilk's Changi, paysheet, 1942-3.

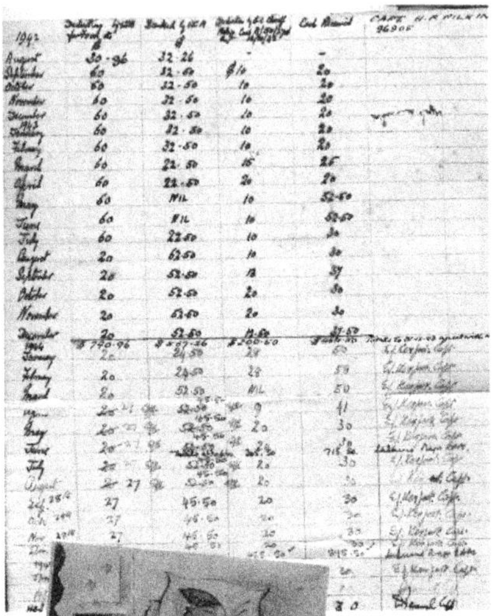

CAPT H.R. PILK...
96905

1942				
August	30-96	32-26	-	-
September	60	32-50	$10	20
October	60	32-50	10	20
November	60	32-50	10	20
December	60	32-50	10	20
1943 January	60	32-50	10	20
February	60	32-50	10	20
March	60	22-50	15	25
April	60	22-50	20	20
May	60	NIL	10	52-50
June	60	NIL	10	52-50
July	60	22-50	10	30
August	20	62-50	10	30
September	20	52-50	13	37
October	20	52-50	20	30
November	20	52-50	20	30
December	20	52-50	17-50	37-50
	$740-96	$507-26	$200-60	[illegible]

Below: A selection of cards received by Dick Lee as a POW.

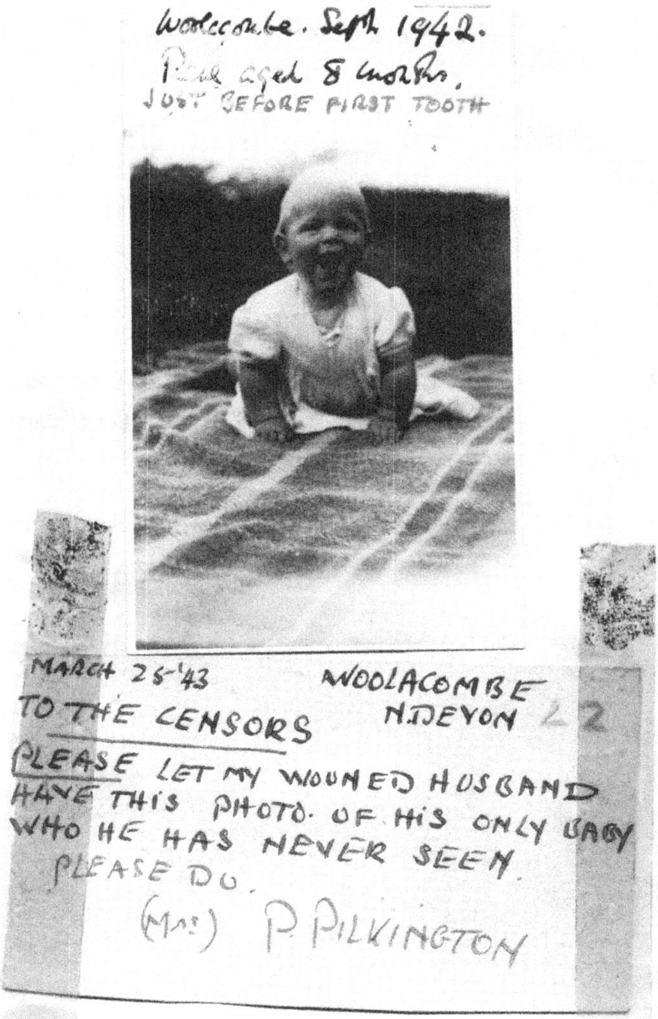

Woolacombe. Sept 1942.
[illegible] aged 8 months.
JUST BEFORE FIRST TOOTH

MARCH 25 '43
WOOLACOMBE
N. DEVON

TO THE CENSORS

PLEASE LET MY WOUNED HUSBAND HAVE THIS PHOTO. OF HIS ONLY BABY WHO HE HAS NEVER SEEN.
PLEASE DO.

(Mrs) P. PILKINGTON

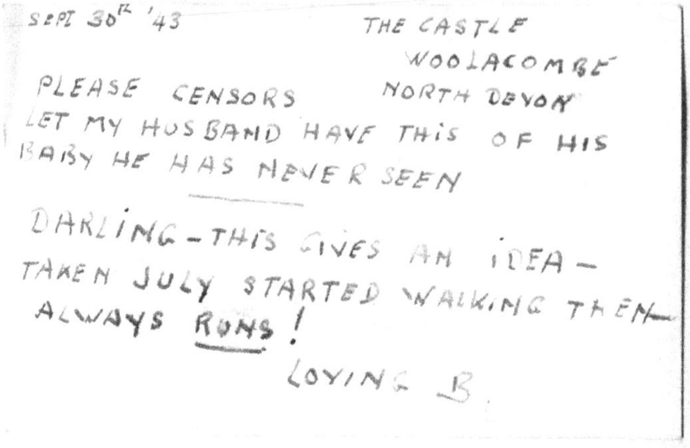
SEPT 30th '43

THE CASTLE
WOOLACOMBE
NORTH DEVON

PLEASE CENSORS
LET MY HUSBAND HAVE THIS OF HIS
BABY HE HAS NEVER SEEN

DARLING - THIS GIVES AN IDEA -
TAKEN JULY STARTED WALKING THEN
ALWAYS RUNS!

LOVING B.

Baby Paul's bare bum was considered a national security threat in 1943! His mum's heartfelt notes to the heartless censor didn't help her cards getting through. (Courtesy of Pilkington family)

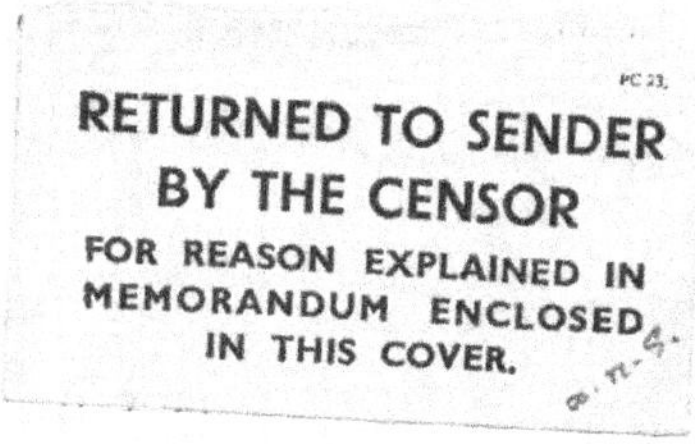
P.C. 23.

RETURNED TO SENDER
BY THE CENSOR
FOR REASON EXPLAINED IN
MEMORANDUM ENCLOSED
IN THIS COVER.

9 ALIVE AND WELL

The Padang, Singapore, 21 December–30 March 1943

'The Padang as it was called was the old garrison playing field. It was a magnificent stretch of levelled grass capable of accommodating three football pitches or two cricket pitches.'

'*Padang*' is the generic Malay word for field. (This padang camp of which Pilk writes should not be confused with *the* Padang at the Singapore Cricket Club.) This was located a few hundred metres north-east of the main Changi Gaol along Changi Road, beyond 'Half Moon Street' which was occupied by the Japanese.

'At one end was a large brick and concrete pavilion with running water. Just behind this playing field were some 50 acres of coconut palms and sand pits with a well drained swamp running through, flanking which were unit vegetable gardens. A number of *atap* huts had been built here to accommodate the Dutch POWs passing through the camp on their way to the Rangoon Railway camps.'

About 15 December there had been a certain amount of rioting in Singapore which had disturbed the calm of the Jap Higher Command who ordered that the 10, 000 British and Australian POWs who were in camps there should be immediately returned to the Changi camp. These prisoners had been used as dock labourers and navvies in general.

'On 20 December I, with other OC units, was summoned to Divisional HQ where the position was explained to us. It meant that we should now have to find accommodation for a further 4000 and, in the circumstances, a complete reorganisation was essential. I was then allocated the so-called Padang area for the 6 Norfolks with, under my command, the 2 Cambridgeshires. The men were to move into the new area as soon as possible, the officers into the pavilion when it was cleared of the tents stored there.

On the morning of 21 December Lieutenant Colonel ————

[name omitted by Pilk] ordered me to move everything including officers mess by 8 pm the following night. I objected strongly and was told it was an order from Division. That night I moved my 100 odd men into the dilapidated huts, still unclean and verminous from their Dutch occupants, but, by nightfall, most of the area immediately surrounding the living quarters had been cleared of all the old filthy cans, bits of clothing, broken coconut shells, etc. and the men tolerably comfortable. All night I kept coconut husk fires burning to keep the sandflies at bay.

The following morning I complained to the RASC that they had not yet cleared the pavilion of the tents. I was told that that was impossible till they knew how the tents were to be allocated. I therefore went to Division and got authority to draw marquees and tents for my own requirements, with an order that the pavilion was to be cleared immediately. Still the RASC refused to take any action. I reported the matter to Colonel ––- [name omitted] who still insisted that I must vacate all my old accommodation by nightfall. In the circumstances I could do nothing but order the immediate transfer of the officers' mess, no mean task as there was no kitchen at the pavilion and everything had to be dragged by handcarts and trailers, a distance of half a mile. Further, all the hens, hen runs, ducks and duck runs had to be moved complete with wire netting, etc.

Thoroughly furious at the way we were being treated I helped myself to all the tentage I required and threw all the rest out on to the grass. This allowed our kit to be under cover for the night and would give some cramped accommodation if it rained and sleeping out became impossible.

In the middle of all this Colonel Dillon and Colonel Hutchinson of Divisional HQ came down and asked me why we were moving in such a hurry. I didn't mince matters and told them that Colonel ––- refused to allow us time to arrange a move in comfort. They were both furious as they had laid down at the conference the previous day that I was to be allowed five days for the move. Colonel ––– was summoned with the result that he received an imperial rocket and was ordered not to interfere with me any more. Further I was told that I could move in my own time. It was now, however, too late as half the gear was down and the other half on its way. It now transpired that Colonel ––- had insisted on our being moved from Quadrant Road, for which there was no real necessity, because he wanted our two houses with their well- stocked gardens for his own unit's use.

More furious than ever I ordered every blade of green vegetable to be harvested, dug up every sweet potato, tapioca root and onions, cleared every papaya tree of even its smallest fruit. The bananas, both ripe and unripe, were gathered for vegetable purposes. I denuded the gardens of everything.

As regards the houses, I removed every stick of furniture, electric light bulbs, ovens and the kitchen store. Colonel ––- was livid with rage and complained to Division. What they said to him I don't know but I was never asked to return a thing!

The move to the Padang having been started, Colonels Dillon and Hutchinson gave me every assistance. The REs built a magnificent cookhouse with ovens and special preparation rooms. Storerooms were built, the roofs of the huts repaired and the sides *atap*-ped, making them very cosy. Shower baths, direct to the water mains, were built and I was given the best of tents and marquees. I was also permitted to draw 200 steel lockers for my men to use as beds and storage cupboards. Electric light was fitted to the sergeants' mess, the orderly room and kitchen, as well as the officers' mess, now established in the pavilion.

A good cookhouse was built for us and I was given sufficient 120 lb tents (large Indian ones) to allocate one to each officer of field rank, one to Jacko, our doctor, and one to myself. The rest were put two to a tent. The batmen were six to a tent and all were soon very comfortably installed.

To give you an idea of the size of the tents I was able to have my bed, chair, washbasin and dressing table in one half while in the other half I had my writing desk, stores and office and still had ample room.

By Christmas Day everyone was content and comfortable and beginning to be very fond of their new quarters. The fowls and ducks I ordered to be shut up with the Dutch prisoners in their detention barracks, an excellent little concrete building that had housed a wireless station and was surrounded by a six feet steel fence. Here the Dutch prisoners, usually undergoing minor sentences for theft, looked after them for us. Later I had to remove them but, by then, proper fowl and duck runs had been constructed near the pavilion where we could keep a watchful eye on them. They gave us endless amusements.

Christmas Day was a whole holiday for everyone. We had all saved up bits of special delicacies from the Red Cross supplies which were supplemented by additional rations issued from canteen profits, CMF and Divisional reserves. Breakfast we had two eggs each. For lunch a curry tiffin, and in the evening a six course meal of tomato soup, tinned fish on toast, ducks (we killed six) and green peas and sweet potatoes. An entree of pig ragout from our share of the pig farm, tinned fruit and dessert – pineapples and bananas.

I had been over to see Mike Louth and found him almost stark naked in his kitchen with a homemade still at work. It was an amazing sight: locked doors, two officers, Jacko and myself, catching the gin as it was distilled in small glasses while sweet potatoes baked in the oven. Funnier still when we were all very merry and opened the oven to get the potatoes, six huge cockroaches walked out in single file none the worse for their baking! Mike sold me six bottles of gin for $5 per bottle. It was excellent stuff and tasted just like Bols. The result was that our Christmas party — to which we invited four or five guests — was a very merry one. I asked two of the doctors who had been so good to me as my guests.'

Doubtless, much chatter would have centred on recent whispers that the Germans had been cleared from Africa by Montgomery and the Second Front was about to be launched. More pertinently, the Japanese were reportedly being blown backwards by the Yanks in the Pacific; and scuttlebutt had it that the Allies were winning back the jungles of Burma. It was all good news that Christmas.

At home, the strains of Irving Berlin's 'White Christmas', with its sense of longing, were being enjoyed for the first time. In Singapore, things were also festive with the re-opening of Happy World, New World and the new Beauty World (not far from the scene of Bukit Timah battlefield) theme parks in Singapore adding some levity to an otherwise austere life. Part of the attraction was that they were off- limits to Japanese soldiers, so Chinese opera and the guilty pleasures of taxi dancers and cabarets could be enjoyed by the mainly male crowds. (Women were largely afraid to go out after dark in case of rapacious Japanese.)

'For the men's meal I had saved up some of their Red Cross food from the boat from which Sergeant Melton produced a most attractive and satisfying meal which they all enjoyed. I was also able to authorise a satisfactory issue of cheroots, peanuts, bananas, etc. to every man, bought with the profits from our own canteen. All the men thoroughly enjoyed themselves and sent along a deputation to thank us for what little we had done and to wish us all a happy Christmas, which was nice of them.'

A letter dated 26 December 1942 from the War Office (Cas. P.W.) to Mrs. HP Pilkington, Ilfracombe read: 'Dear Madam, In confirmation of War Office telegram dated 15 December 1942, I am directed to state that information has been received that Captain HP Pilkington, The Royal Norfolk Regiment, previously reported missing, is now a prisoner of war in Japanese hands. It is stated that he is interned in "Malaya Camps" and although it is not yet known whether this is a complete address, you are recommended to use it as such in communications to him. Any further information will, of course, be sent to you immediately. It would be appreciated if you could notify this department of any information which may come into your possession concerning your husband's location, welfare, etc.

Information regarding the means of communication with Prisoners of War in Japanese occupied territories may be obtained from your local post office. I am, sir/madam, your obedient servant [A.S. Weston]'

'On 26 December the troops from Singapore started to roll in and I found that I was soon catering for another 300 men, 6 Norfolk, 2 Cambs, and other units. They quickly settled in and we settled down to improve our quarters.

The first three days every man was employed collecting all the scrap iron and carting it away (some 200 loads of about one ton each). All the scraps of clothing were collected and burnt and all the coconut husks lying about over the 50 acres were carted in and dumped by the kitchen for fuel. The results were most satisfactory.

Mosquitos and sandflies literally disappeared, the place smelt sweet once again and, when large grass-cutting parties supplied by Division had cut down all high grass, very tidy and pleasant.

A barber's shop was built with a real barber's chair looted from Singapore, a large recreation hut fitted with tables, forms and electric lights constructed and water carts filled with chlorinated drinking water stationed at strategic points. Salt by now was worth its weight in gold so I had salt boilers built which were fired by the coconut husks, the water being fetched by special authority every day from the sea by the unfit men for whom they brought up in mule tanks some 200 gallons of salt water daily. Four and a half gallons of sea water gave one pound of salt so it was now possible not only to improve the cooking beyond all recognition, but to make a salt issue to the men for their own use.

Daykin took over the canteen and made a wonderful success of it. Amongst other things we cornered forty-eight tins of beef dripping and, as at the same time bread was starting to be manufactured from rice and maize flour, we were able to sell the men twice a week a huge hunk of bread, dripping and salt for 20 cents. They loved it and always looked forward to it. Needless to say the officers had it too. Ducks eggs were now obtainable at 10 cents each and very good they were. I spent most of my money on them, giving Carlton two a week which, as I had insisted that our batmen should share all our eggs, did wonders in keeping them all fit. It was agreed that our batmen, who were also our friends, should share all our food, etc., a decision I don't think that any of us ever regretted.'

But not everyone was enjoying the same good conditions. By end of 1942, 29 civilians had died in Changi Gaol. But the British also proved themselves cruel gaolers: 106 Japanese internees (among them 42 women and nine children) held by the British in Purana Quila camp near Delhi, died from exposure to extreme weather, others from malnutrition, especially beri-beri, but mainly dysentery.

'With so many troops once again with us, all the big and little theatres started up again. Besides the New Windmill, the Hippodrome and Hospital Palladium were opened. Both of these constructed or converted from the garrison cinemas. Amongst other shows put on at these theatres were Sheriff's *Badgers Green,* an excellent play, *Jack and the Beanstalk*, a locally-written pantomime done in a Norfolk accent with Giant Nip (our term for the Jap) as the centre character surrounded by his captive prisoners of war. It was excellent, particularly some of the ladies that must have scaled nearly half a ton!

Journey's End, a poor production and quite out of place, *Ten Minutes Alibi*, a thriller very well produced and acted and, finally, a really super show called *Dancing Tears*, locally written with stage effects and scenery that would have shamed most provincial theatres at home.

But what I liked more than all and never got tired of were the symphony orchestras and concerts in which musicians, violinists and singers that were at one time household names on the BBC took part.

There were, of course, too many small theatres constructed by the men for their own use, such as one known as the Old Water Wheel where they had gone to infinite pains to overcome all difficulties. The ground had been sloped to give all the audience an uninterrupted view, the orchestra operated below the stage in a pit dug out for the purpose, and the scenery and effects were most ingenious. All the clothes were made out of old bits of mosquito netting and tenting dyed to suitable colours by homemade dyes, tinsel and tissue paper filled many gaps, while some most lovely wigs were constructed out of coconut fibre and the stuffing from mattresses. All these shows were written by the men themselves and produced without any assistance. If somewhat crude and vulgar they certainly gave us many a good laugh.

The big theatres were supplied with all they required in the nature of paints, grease paints, musical instruments etc. by camp funds, while every party that went to Singapore swelled the actors' wardrobes with looted ladies' clothing as well as men's. The result was that they could turn men and women out in immaculate evening clothes to the most fantastic costumes.

In the Australian area was a huge theatre run by a proper Australian concert party that had been with them at the capitulation. Their female impersonators were perfect, while their drill, lighting effects and sketches were masterpieces. My Australian friends invited me regularly to their shows which I always found well worth the three mile walk.

Organised games, football and cricket matches were again started and of these from our pavilion we had front line seats. On Sundays there were first-class cricket matches, including a series of Test matches between England and Australia. Unfortunately the pitch was badly cut up so a coconut mat had to be used, but they were a very pleasant sight, the players whenever possible wearing white flannels which gave them a very home-like appearance. [Australian Test wicket-keeper Ben Barnett was among the players who turned out.]

During this period large working parties were supplied daily for the unit and Jap gardens, as well as the forestry parties. I received a very nice pat on the back for the way I ran my area, the biggest in the camp, which I did as if I was running an estate. All this time I was unable to read, my eyes getting so bad and inflamed that I was only allowed out in the early morning and evening, except to go down to my kitchen as I had retained the unit feeding in my own hands.

At first there was a good deal of friction between the 6 Norfolks, 2 Cambridgeshires and those that had returned from Singapore. We had Red Cross food left, others had eaten all theirs; some had a lot of money but no food, others had none. It took a lot of sorting out but by gentle hints I eventually made them all come to a satisfactory agreement whereby differences in stocks were made up by cash payments. After this, all went smoothly and I must say I was very touched when we finally left this area when a composite deputation of all the units under my administration came up and thanked me for all I had done for them.

On Sunday evenings we arranged talks for them by officers recently captured by raiders or shot down in the air, all of which were much appreciated.

On New Year's Day the Japanese had made us all a present of 1/10th of a bottle of sake to drink to their Emperor's health. It had, however, a slight kick in it.'

With the New Year – ushered in with the singing of *Auld Lang Syne*, which had escaped proscription by the Japanese unlike a thousand other American and British songs – absolutely nothing was known of the location or condition of camps and prisoners on the Railway, to where Dunlop Force (which had originally sailed from Java) was currently heading.

On 7 January 1943 *The Times* ran a piece headlined: **'Prisoners in the Far East. 10, 000 now in Japan. Severe conditions in Malaya'**: 'The War Office states that information about prisoners of war in the Far East is still fragmentary and sometimes conflicting. The Protecting Power has received permission to visit certain camps in Japan. [For POWs in Japan] The standard of living to which our prisoners of war have had to adapt themselves is that of Japanese troops. The food consists largely of rice, though some bread is issued and fair quantities of vegetables and fish. The meat ration is extremely small. The majority of United Kingdom and Australian prisoners of war are still in camps in Malaya, Siam and Indo-China. No visits by the Red Cross or by the Protecting Power have been permitted. Work in these areas is understood to be very severe. Rations appear to be just sufficient to maintain health.'

At home, Phyllis was clearly still fretful, penning letters of enquiry trying to find out where the father of her one-year-old boy was. On 16 February she received a reply from the Vatican War Enquiry Department, Cavendish Square, London, written in a flowing hand: 'Dear Mrs. Pilkington, A reply to the enquiry you made some months ago has just come through on the Vatican Radio. I am happy to tell you that Capt. Hugh Pilkington is safe in a POW camp in Malaya. I congratulate you on knowing he is safe. You may have heard this already from another source.'

For once, the information was accurate at the time. On 20 February a further letter was received from the Vatican: 'Dear Mrs. Pilkington, I am afraid we have no 'inside' information about camps in Malaya. The men appear to be treated quite decently although the food standard is low for Europeans. You can send a 25-words message through us. It MAY reach him so it is worth trying. Write it on any sheet of note paper — we have to copy all messages onto special forms (charge = 2 1⁄2 stamp).'

Phyllis was also writing to Pilk regularly, although the postal schedule was less than optimal. On 6 March 40 mailbags (representing a large part of the approximately 250, 000 letters to reach Singapore that year) arrived, following a nine-month roundabout route from London.

On 25 March Phyllis sent a photo of Paul lying down smiling on a picnic blanket at Woolacombe: 'Paul aged 8 months just before first tooth'. On the reverse: 'To the Censors: PLEASE let my wounded husband have this photo of his only baby who he has never seen. Please do. (Mrs.) P. Pilkington.'

While Franklin D Roosevelt called for a US$100 million military budget (at the same time introducing coupon books in a new processed food rationing plan), 1943 was largely characterised by a stalemate of sorts between the Axis and the Allies, and very few changes in the map of the Pacific resulted. South-East Asia was not really a priority for the Allies, pre-occupied as they were with building up to the invasion of Europe. Japan, too, was unsure of how next to tackle China with the US fleet making headway through Guadalcanal and across the Pacific.

The rate of Japanese ships being sunk increased dramatically, largely thanks to deadly submarines prowling the narrows that Japanese shipping was restricted to. This of course made the need to construct the Thai–Burma Railway link even more pressing.

In the middle of March the Japanese started sending more working parties up to Siam to work on the Railway, and, we have since learnt, to replace those that had died or were too sick to work.

This was D Force, 5000 men departing Singapore 17 March.

'The result was that by the end of March our numbers had sunk considerably, so much so that the Japanese ordered that all temporary buildings and tentage were to be evacuated and everyone returned to buildings, so back we had to go to Quadrant Road, a very small party by now.

Without doubt this was the happiest period of my life as a POW.

There was something about the atmosphere amongst the coconut trees that stopped the eternal grousing and squabbling; everyone seemed happy and it was only with the greatest regrets that we left. Since then many officers and men have told me how they appreciated this period, in particular as they were all treated alike, everything was shared evenly and all complaints promptly dealt with.

A small matter that often pleases me to look back on was that every time a coconut fell there would be a mad rush for it, the race always being won by some man who was supposed to be incapable of leaving his bed.

A great deal of the contentment amongst the men was undoubtedly due to the humane and kind medical treatment given to the men by Jacko. He deserves the greatest credit.'

10 SAYONARA SYONANTO

Quadrant Road, and Garden and Woods, Singapore, 30 March–10 May 1943

'In all in the last moves, although I had lost no officers, I had lost every one of my fit men except three batmen, including Carlton, leaving me with about 120 men and two or three sergeants. From the mess I had lost some of the attached officers such as Major Galloran, RAMC, and some of the 2 Cambridgeshires.

By now all Red Cross supplies from the ship that came in September were exhausted and we had to fall back on whatever we could buy or produce ourselves. The Japanese were now allowing us 15 oz of rice a day, some seven grams of salt, five grams of dried fish, a little tea and practically nothing else except masses of peppercorns.

We returned to our same building in Quadrant Road except that we now occupied houses 2 and 3 instead of 1 and 2. Everything was just the same as before; our neighbours comprised the rest of the Infantry officers of the Division less those of 5 Norfolks who were living in the detention barracks with Colonel Prattley. It was now very much pleasanter than before as Colonel ——- [name deleted by Pilkington], through no choice of his own, had been sent up- country, much to everyone's satisfaction.

At the beginning of April I received six letters giving me the news of Paul's safe arrival. Although I was quite certain of this it was a great relief to know that they were both well.

Mike Louth gave a party for me to celebrate this news and we all, much to the envy of everyone, got really happy on his gin, so much so that Hugo Hughes – who was by now out of hospital with only one leg, on crutches – danced all the way home. The pineapple gin had given out early in the evening and we had then regaled ourselves with *gula malacca* beer which Mike described as 'three days off full strength'. What its full strength was I never learnt as I gathered it had never been allowed to mature that length of time, but it was certainly a powerful beverage.'

By now, up-country in Thailand, heavy lashing rains signalled the start of the dreaded summer monsoon period. Worse, the Japanese scaled up the workload on the railway to meet their ridiculous new deadline of August 1943 rather than the equally unfeasible December. This was called the 'Speedo Period' for the insistent shouts of 'Speedo! Speedo!' from the Japanese guards as they coerced the workers, verbally and physically, to add to the mere 20 percent of railway line completed by that time.

'In the middle of April the Japanese ordered two more working parties: one for Borneo of 1500, and one for the Rangoon Railway of 7000, known as F Force. All my fit men having already gone, I only had to produce six men for Borneo.'

Out of B and E Forces totalling around 2700 men, 1053 ended up on the infamous Sandakan Death Marches. Just six, who escaped, survived the war.

'With this last party to Siam went Mackwood as one of their concert party and Padre Duckworth of 2 Cambs who had been with our Brigade throughout the war. When this force left they were told by the Japs that they were going to be a wonderful new camp in the hills and that they were to take everything they required. The result was that a special train load of baggage moved off consisting of everything from a field bakery, an electric lighting plant and organ to the whole of the Divisional library, their share of the food reserves and even their chickens. The last they saw of all this was in the railway sheds at Ban Pong being distributed to various Jap units.'

Ban Pong was the railhead in Siam initially, and now around the '5 kilometre' marker west of Nong Pladuk for the new railway to Burma.

When F Force had gone there were only some 15, 000 British and Australian troops left in the Changi camp, so the Japs ordered the closing down of 18 Divisional and Southern areas. The latter was eventually used for their own troops while the former was turned into a huge aerodrome which covered the Padang, pavilion and all the coconut groves and stretched for some three miles by one.

'About this time the Japs announced that telegrams to relations in Australia would be permitted. I therefore sent one to my

Uncle Harold asking him to tell Phyllis I was all right. I don't suppose she ever got it from him. If it had been transmitted I am sure the Australian Red Cross would have sent it on to its destination.'

Australia by this stage only had official news of some 3000 of its 21, 000 POWs. In the greater South-East Asian theatre, 250, 000 Allied troops were missing and, as yet, untraced or unreported by the Japanese.

'An influx of so-called Dutchmen, but in fact nothing but half-caste citizens, had brought with it a lot of native cooks, who now taught us what really could be done with rice, rice flour, coconuts, *gula malacca* and other native foods that could be purchased and there can be no doubt that the knowledge they brought was worth its weight in gold, as all cooking which had been monotonously British, now took on a new sparkle and excellent new dishes appeared, attractive both in looks and in flavour.

A proper central shop was also started where little luxuries could be purchased such as razor blades, toothbrushes, combs, etc., and where also those requiring money could sell things they didn't require. I was now able to buy myself a large and good mosquito net for $2/50 that served me faithfully to my last night in Changi Prison, as well as a brand new pair of crepe soled walking shoes for $6 which I only discarded on my last day as a prisoner of war, but by this time they were kept together by yards of wire but they were still serviceable.

About this time too we were allowed to send off our second postcard of 24 words.'

This does not exist among Pilk's comprehensive memorabilia, although he was led to believe she had received it.

'The general tone and morale throughout the camp was by now excellent as the successful North African battle was drawing to a close and we all felt that before long an effort would be taken to recapture Malaya and us.

On 20 April we moved our household goods by hand trailers to the Gardens and Woods area, a distance of some 2 1/2 miles. I left nothing behind for the Japs, every stick of furniture and all fittings being taken. I did however insist that our gardens, in particular my brinjal orchards which were by now yielding well, should not be despoiled but left for Roberts Hospital. Many doctors and people have since told me what a godsend these vegetables and fruit trees were to them as, being close by, they were able to harvest them with care.'

In May 1943, the Japanese took over Roberts Barracks, needed for their airbase personnel.

'The gardens and woods area which was now to accommodate the whole of the remains of 18 Division was a series of some fourteen large huts of the wood and *atap* variety.

On the whole we all fitted in very comfortably. I was allotted a little room to myself into which I got all my belongings, including a long-padded chair bequeathed to me by the Brigadier. The place was very hot, full of mosquitoes, with a very depressing outlook on to some ghastly old rubber trees. Modern sanitation which prevailed generally elsewhere was here non-existent and had to be replaced by latrines of the bore-hole and trench type, which always smell abominably. The water supply was very poor and the pressure weak. On the whole I was depressed with the area from the very start.

The whole area only contained three cookhouses; these were allotted two to the men and one to the officers. The total number of officers was around 80, which meant a terrible squash at meals and precluded the use of mess waiters, which meant that we had to be served on the cafeteria system; queuing up for each course, a loathsome proceeding. Furthermore one had to take into each meal one's own plates, knives, forks and mugs and bring them out dirty afterwards for Carlton to wash up.

The food here was very indifferent and badly cooked. I cannot recollect one decent meal. Fortunately I had divided out amongst our old mess members our surplus tea, coffee and fruit so we were able to make ourselves an evening brew. Eggs too were still plentiful and, with the depleted numbers, garden produce abundant.

The only good thing that can be said for this period was that Mike Lough, Shotter and Hugo were now near neighbours, they also having had to move into the adjoining area, which had previously been entirely Australian.

The Australian theatre put on some excellent concerts, all –

both songs, words and sketches — written locally. As my eyes were troubling me considerably I was not allowed to do any work in the sun.

Early in May the Japs ordered another party of 3000 to be ready to go to Thailand (Siam). They stated that they were fully aware that nearly every fit man had already been sent, but assured us that this party would be for administrative duties in a healthy area. Further they strongly advised that all those suffering from deficiency diseases, ie, beri-beri, painful feet, eyes, etc. should be sent as the food there was plentiful and good. They were so insistent on this that our medical authorities could do nothing but concur, and anyhow it didn't matter if they did not because the 3000 men would have to be found in any case.

As my medical category was 'Permanently Unfit' I was naturally not included, but Carlton and twenty-four other of my semi-fit men were. I could perhaps have pulled more strings to get Carlton off but it has never been to me the right thing to make a sick man go in the place of a fit one because of an unfair wangle. At the same time I myself was willing to do anything to get my eyes right which would only be possible with good food. The risk, in spite of the fact that I didn't believe a word the Japs said, appeared to me justifiable. The surgical specialist agreed to my going providing I went in an administrative capacity and did not attempt to carry heavy loads.

After due consideration, taking into account the fact of my eyes and that I disliked sending my own men off without one of their own officers, I asked to be included.

At first my application was refused but later, at the request of Monty Smyth who was to be in charge of the party, it was accepted. I therefore went up as second in command of the 18 Divisional unit. In spite of all that I had to go through and suffered, it is a decision that I have never regretted and the fact that my eyes are nearly fully recovered is I am sure entirely due to this excursion.

Fortunately Jacko, our MO, was also detailed as MO of our party so that I was always under his kind supervision, which meant a great deal.

This party was eventually detailed to move off on May 10.'

History was to know this party as H Force.

11 LIKE PIGS TO MARKET

Journey from Singapore to Thailand, 10–21 May 1943

'The decision made, it was necessary to hand over the affairs of the Battalion. These consisted almost entirely of the records of killed, wounded and missing, which we had compiled from all available sources over the last year. These I gave to ——- [name omitted by Pilkington].

With profits from our unit canteen and subscriptions from the men out of their pay, I had, with Daykin's help, amassed a large reserve supply of tinned fish, bully, margarine, M&B, etc., not so much against possible starvation, but as an investment. Tinned foods were becoming scarce and every month now a rise in their prices. Many, like dripping, were already unprocurable.

As each party had gone up-country I had given each man his full share which amounted to at least one tin of fish and $1 each per head. I now asked Daykin to make a final distribution for those 25 men going with me. The balance left over amounting almost entirely of tinned food as I had insisted on taking the cash rather than the food as we would, it seemed, be going to a land of plenty. This tinned food I left with ——- in trust for the men to be sold or distributed at his discretion.

In order to square up the mess accounts it was necessary to distribute the hens and ducks. My share amounted to three hens and my muscovy duck, Olga, worth in all some $12. In order not to break up the farmyard I agreed to sell these birds back at $6 provided I and Carlton were each given two extra eggs for the journey. This was agreed. To my horror and rage when we were packing up and I had received my two eggs I found that ——- had refused to allow two eggs for Carlton. My threat that I was going to report the matter immediately to higher authorities produced the eggs in a hurry, but for downright meanness I have never met such a case before or since.

As, with my damaged shoulder, I was unable to carry a pack,

Carlton constructed four little canvas panniers with broad canvas straps that hung over my shoulder back and front and into these I was able to pack all the essentials of life, while my cape and blanket hung over the shoulder. We all took one book. I took the large edition of H.G. Wells' *History of the World*. I also carried a small haversack and a water bottle.

The rest of my clothing, blankets, and kit I split evenly into two piles so that if one lot got lost and the other turned up I should still have a proportion of the essential goods. These were to travel as luggage.

I bequeathed my furniture to various friends as I never expected to come back to Changi; my bed and mattress being a particularly valuable gift. I was asked to a round of coffee parties, all very enjoyable.

Stalin the muscovy duck had supplied us with our farewell meal at Changi; stuffed with sweet potatoes and herbs, roasted to a turn and served with an excellent burnt sugar gravy, he had filled six officers and three batmen, including Carlton, to repletion. Truly a noble bird. He was about the best duck I have ever tasted, and what a lot of breast meat.

All was now ready for our departure. On the night of 9 May we marched down to the Changi Road where we were checked through the gate, name by name, and we then fell in on the road in parties of 25 ready for the trucks to arrive, which they did at 1 am. The officers filed into the front truck, there were final goodbyes, a whistle was blown and we were off.

This was the end of my first term of imprisonment as a POW. Up to now very few people in Changi had come into contact with the Japanese — a large number of people I don't think at this stage had even seen a Jap — as the few guards we saw were Koreans, a very different matter. We had administered ourselves, distributed our rations, and generally behaved more as a unit in the field than a force of men that had capitulated to and were under an alien power. All the majority of people had heard were stories of Japanese brutality and barbarianism that they had refused in their minds to contemplate as being true, in spite of daily evidence brought in by working parties.

I saw Changi recede with mixed feelings. Undoubtedly it had

been a comfortable existence, we had been unmolested. Thanks to our gardens and central messing funds we had had all we required to eat and we had, within the wire, enjoyed absolute freedom. Against this was the fact that nearly all my friends who had gone up-country, my eyes were aggravating me badly, and the community living that was latterly in force I found very trying. On the whole I was glad to be on the move.'

H Force had a total of 3000 men. For travelling and administrative purposes the force was divided into five parties of 600 each. Each 600 comprised a train-load and carried its full administrative branch — doctors and RAMC orderlies — bringing the total for each party to 650.

'The drive to Singapore in the dead of night with a full moon shining I found exhilarating. It was the first time I had been anywhere except in Changi or in hospital since I was wounded.

Singapore was like a city of the dead. There was none of the night time clatter with street hawkers sitting by their stalls with little lamps. The place was deserted, and over it one could feel an ominous hatred of the Japanese.

We arrived at the main railway station where we debussed and marched to our trucks, the train having been drawn up at the main departure platform, with the railway hotel where we spent our last night in Singapore before sailing on the *Gneisenau*, deserted. [On their final passage home in 1937, Pilk and Phyllis had sailed on the German ship that was famously converted to an auxiliary carrier in 1942.]

The only meal I have disliked is sauerkraut and German sausage, which we had on the boat home.'

Pilk would have quietly appreciated the 1932 station adorned with Federated Malay States heraldry, a result of the sale of Singapore Railway's assets to the FMS for $4 million in 1918. Although the building had been bombed by the Japanese, there were beautiful mosaic murals of bucolic Asian scenes, including rubber tapping. Initially the railway line terminated at the Johor Straits, with two boats ferrying passengers across to meet the line on the other side. With the completion of the Causeway in 1923, trains could finally cross from Singapore island to the peninsula.

'The train consisted of about 24 trucks of the steel variety into which we used to load rubber for transport to Singapore. Beyond their steel sliding doors they had no ventilation and there was no straw on the floor – just steel. Into each truck had to be squeezed 27 men and their personal belongings, with wooden table tops placed across the door openings to prevent anyone falling out. This allowed per man just sufficient space to sit on his kit with his knees drawn up. The railway journey lasted five days.'

Long delays en route were not uncommon as the POW trains were considered low priority, sitting in sidings while other more important cargoes of troops and supplies took precedence.

'I was more fortunate as Monty, myself and Jacko were travelling in a truck with the four Jap guards. The Japs took one half and we the other, which gave us ample room to store our stuff and lay out bedding roll at full length.

Before entering the trucks the Japs had a 'Tenko' or roll call when we were checked truck by truck. As soon as everyone had got in and settled down they made everyone get out for another one. Just sheer devilry, as they found it amusing to see the men cursing and swearing when they tried to settle in again. At last even they appeared satisfied. It was now about 5 am and very cold on the steel floors but, nevertheless, I was dog-tired so I laid out my bed and went to sleep.

At 6:40 am we were woken by the engine shunting on to us, when all the trucks cannoned into each other, when, if you are not careful, you will lose your balance or get badly bruised. A few minutes later a whistle blew and we were off. It was still pitch-dark and I went on sleeping for another hour.'

Had Pilk been awake, he would have seen the rear of the Alexandra Hospital from where the Japanese had attacked fifteen months earlier.

'When I next looked out of the open doors we were crossing the Causeway into Johor.'

Railways in Malaya were originally built for tin-ore transportation in colonial Perak. Running between Port Weld (now Kuala Sepetang) and Taiping, eight miles of track was built by the British in 1885. This journey took around 40 minutes. The Seremban–Sungei Ujong line followed in 1891 to link it with Port Dickson. The ensuing rubber and tin boom necessitated, and financed, an efficient rail system that joined all the individual pockets of lines.

In 1901 FMS Railways was founded, merging the previously state- run operations. Two years later, Pilk's adopted town of Seremban was linked to the federal capital, Kuala Lumpur. Soon, a continuous service was available from Seremban up to Prai, near the border with Siam. And, significantly, the East of Malaya was opened up with a line from Gemas (in southern Negri Sembilan) forking eastward to Khota Bahru in 1931. Opened by Sir Cecil Clementi, this line braved wild and rugged scenery, predating roads in many parts of the north-east coast. Indeed there was only one road over the mountains to the east: rivers were the high roads.

'As the sun got up the trucks became like ovens; the steel walls got so hot that to rest one's head on them was impossible. The men packed tighter than sheep in a train at home, suffered agonies.

Every now and again the train would stop when, not waiting for orders, everyone leaped from the trucks and made a rush for the trees as there were no sanitary arrangements in the trucks. At wayside stations it was possible to buy fruit, papayas, pineapples and bananas comparatively cheap, the natives in many cases giving their fruit away to those who had no money. Our train halted beside a huge pile of bananas to which our Jap guards told us to help ourselves. We did and the clump disappeared but, as the train disappeared around a corner, the real owner turned up hurling curses on us.

All day I sat on my little canvas chair that Griffiths had made for me, watching the world I had known in different circumstances go by. Rubber estates were unworked, but essential upkeep was still going on. Tin mines were nearly all abandoned and flooded. An air of poverty, depression and hatred hung over the country.

On Gemas station we were given a meal of rice and weak bean stew with a mug of hot water to drink. As meals went it wasn't too bad.

We passed through Seremban at 16:00 hours but only stopped for a few minutes. I was, however, able to learn that on the estates food was still available and that there had been no outbreak of diseases. Ainsdale Estate looked very dilapidated and abandoned, but the factory and smoke houses still stood. I caught a fleeting glimpse of our bungalow.'

Had Pilk been able to walk into town, he would have found a new feature on the landscape: Japanese brothels. Girl's House and New Girl's House were operating in town, with *kimono*-clad beauties, one of 100 Japanese 'comfort stations' operating in South-East Asia. One woman for every 40 soldiers was their provision and 32.1 million condoms, mostly 'Assault No. 1 brand', were supplied to Japanese units in Asia.

'We arrived at KL main station platform at 16:45. The platform had been cleared of all local natives and had guards at either end, so that it was impossible to contact anyone. After the usual mad rush for the latrines was over — when every one of them including the Ladies, Porters and Natives were called into use — we were given another meal of rice, watery vegetable stew with a few bits of pig floating round, and tea. At the same time water bottles were filled. I, by luck, ran into the stationmaster who secreted me into his office where I had a quick talk in his private lavatory.

I came across eight Australian officers and men intent on picking out the lice from their clothes. So concentrated was their attention on what they were doing that I asked them why the grim silence? I was informed those finding the greatest number of lice would not have to pay for their morning round of coffee.

At midnight there was another *tenko*. We scrambled aboard and were off again.

The night was hot and stuffy and the carriages refused to cool down. However I slept well and woke to find the train well up north. When leaving KL we had drawn breakfast in our containers which consisted of dried rice with a small piece of dried fish per head. This was eaten while the train was on the move.

Today, 11 May, was a repetition of the previous day. The men cramped and hot in their trucks, made rushes at every stop and good-naturedly haggled with the natives for fruit. We reached Ipoh at 17:00 hours and crossed the Taiping Hills with the setting of the sun, a beautiful and grand scene, reminding me more of pictures of the Rockies than Malaya. That night it was very cold in the trucks.

At 2 am we arrived at Prai, where we were given another hot meal (we had had one at Ipoh) of stew, rice and hot water and we also drew our breakfast of rice and dried fruit. We stopped here for about an hour and I took a stroll along the quay, no one stopping me, and looked at the lights of Penang.'

How Pilk's circumstances had changed. Penang for him previously represented joyous scenes in Georgetown of cruise ships to and from home leave, where the E&O Hotel graciously presided during the Great Age of Travel — Port Out, Starboard Home and all that. Fortified as Penang was by the Royal Engineers at Batu Muang since the 1930s, the Brits had deserted this — their largest integrated fortress and military operations complex on the south-east of the island — fearing rightly that the island would be isolated as the Japanese front swept south down the mainland.

'After four o'clock after innumerable delays, the train left and we woke to find ourselves threading the paddy fields of Kelantan, full of caribous ploughing-in the mud, ducks swimming in the drains, and early natives going to market.

We reached Padang Besar at 9:30 am. This is the border station between Malaya and Thailand. Here the train was shunted on to a siding and the FMSR engine cut off. The driver had been very good to us, giving us all the hot boiling water from his engine we required to make tea.

At Padang Besar the Japs made everyone get out and do twenty minutes PT. We were allowed to sweep out and tidy all trucks after which, those of us that were lucky, had a bath under the water hydrant usually reserved for filling the engine. It was delicious and from now onwards a rush was always made for these water stands which the station officials very good-naturedly kept on for us as long as the train was there.

At 2 o'clock on 11 May an engine belonging to the Thai Railway took us over the border into Siam.'

Traditionally, the people of Siam had relied on elephants, buffalo, and ox-carts for transportation and travel. It was revered visionary King Chulalongkorn (Rama V) who issued a royal proclamation in 1891 that a railway line should be built. Almost three years to the day later, on 26 March 1894, he presided over the opening of the first rail service from Bangkok to Ayutthaya, the ancient capital. The Royal State Railways of Siam was born and by 1910 an impressive 774 kilometres of track were in operation.

The royals kept a hand in the railways, with Prince

Purachatra becoming its first Commissioner-General in 1917, and pushing for modernisation. By 1925, 1804 kilometres of track was in operation, the first diesel engine running on its tracks three years later. His biggest challenge was merging the original Northern and Southern Railways which operated on the east and west banks of the Chao Phraya River, respectively, and on different gauges. The prince insisted on a ten- year conversion program to bring all tracks in line with the gauge that would be uniform with its neighbours: Malaya, Burma, Cambodia. This was duly completed in 1930. By the mid-1930s, 2222 kilometres of track carried traffic.

'All of 12 and 13 May we trailed through this new country. Much to my surprise it was not flat and uninteresting as I had expected but was generally huge expanses of paddy fields which could only just have been above sea level, with huge steep-cliffed granite islands and hills sticking up. It looked as if only a few years ago it had risen from the sea bed, cliffs, beaches and all. The Thai railways are wood-burning and we now required two engines to pull our train instead of one.

In one case in Thailand — where the water had to be pumped from a river — our departure was delayed an hour as we had used all the water in the tank for bathing and most of it out of the engine boiler for tea. Imagine our amusement, and our guards' anger, when it was reported that the train couldn't go as we had used all the water from the engine to make tea.

We travelled up the backbone of the Malay Peninsula till we reached a small Thai town called Ban Pong at 8 o'clock on 14 May. Here our journey ended, we got out of our trucks on to the platform and unloaded our kit.

The railway journey for the men had been a ghastly nightmare of cramp, heat and cold. For me it had been comparatively comfortable.

The Japanese Sergeant in charge of the train was as nearly human as it is possible for a Jap to be. He refused to let his guards beat up the men, gave them every liberty which, providing they did not abuse, gave them the privilege of trading with the natives, getting down at halts and bathing when possible.

In our carriage he was consideration itself, and he and his guards invariably shared all their meals with us, saying that they were sorry that the food was so bad. Their meals were excellent and most tasty, particularly the pickled plums. Another advantage was that they did our shopping for us when, invariably, we got our purchases for practically next to nothing, the most prized being huge hard boiled ducks eggs at five cents each, and legs of curried chicken at 10 cents.

When we parted at Ban Pong the Sergeant gave me his 'housewife' [bolster pillow] as a keepsake. He was good to us.

In spite of their degradation to the level of pigs being taken to market, and to the eternal surprise of the Japanese, the morale of the men was never higher. They cheered their way through every station and sang their days away without apparent thought of their discomfort. It would by any observer not conversant with conditions, have appeared to be the triumphant progress of victorious troops; it would never have occurred to him that they were POWs.

One point that amused me was that all the station masters in Thailand were dressed up in uniforms covered with badges and gold with big peaked caps, very much in the German style. Even their minor officials would have put a staff officer to shame.

Having detrained at the station at Ban Pong, a *tenko* was held, the men were fallen in with all their goods and chattels and marched to a transit camp about 3⁄4 mile away. I was left behind to load the heavy baggage and get a free ride. We arrived at the transit camp which was nothing more than A-shaped *atap* huts 200 feet long and 40 wide, with bare earth floors. Everyone had to unpack their possessions and were then thoroughly searched by the Jap military police, not so much with a view to finding incriminating documents and wirelesses, but with a view to seeing what it would be worthwhile purchasing later.

A tommy-gun with 50 rounds of ammunition, several compasses, a couple of loaded revolvers, and half a dozen wireless sets hidden in water bottles escaped attention.

While our cooks were preparing breakfast with the help of some Australians who had been left behind to help us out, we were allowed to stroll about as we wished.

In Ban Pong we found many of our troops that had now been in Thailand six to nine months. We were able to give them the good news that Tunis had fallen, a bit of information received only an hour before we left Changi. It was grand to see how they all brightened up with the confirmation of the first great victory of the war.

Later in the morning a Jap officer came and said that we were to start marching next night to a railway camp 130 miles up in the jungle. It was explained to him that we were a party of sick men who had been promised that they would not have to march or do hard work till they were fit. He laughed and said he knew nothing about that — all men in Thailand worked, that our place was on the railway, and to that place we would march. Further that we could consider ourselves extremely lucky as the previous party that had left there weeks before us was still marching as they had to go right into Burma, some 200 miles. He did, however, promise to send what kit we couldn't carry on behind us.

It now appeared that Ban Pong was the junction for the new railway and was also the road terminus of the only good metalled road that ran to Kanchanaburi, some fifty miles away [actually 47 kilometres].

Ban Pong was nothing more than a few sturdy stone buildings, although its main street bustled with makeshift stalls of produce vendors.

To look at Ban Pong was much like a miniature Seremban with a central feature of a large and rather lovely Buddhist temple, a mass of bright native colours, clever and well executed plaster and wood carving with a bright green tiled roof decorated with dragons belching fire. A yellow robed priest with shaven head showed me over. It appeared that in the morning and evening it was used for worship, during the daytime as a school, and at night as a brothel! In the temple grounds were hordes of bald-headed vultures sitting on the sacred trees waiting for anything to die. They were an interesting but eerie sight.

Our meals here consisted of three good meat-and-egg stews a day. The stew of pork, water and heaps of vegetables was brought to the boil when about six eggs per head were broken into it. The result, eaten with rice, was first-class. From now onwards all drinking water had to be boiled, a great hardship as marching men require a lot, and a water bottle full a day was all they got.

A near tragedy: Carlton had his water bottle stolen and it was only after hours of search and bartering that I was able to procure for him an ordinary glass bottle as a substitute, by which time he was in almost hysterical despair, with weeks and months of thirsty work and marches staring him in the face.

There was the poor man who had lost his spoon, the only instrument with which one could really eat rice, and who I found broken down in tears at the edge of the jungle, and his pathetic gratitude for a teaspoon I was able to give him as a stop gap.

Around the transit camp — which had a huge notice up 'Rangoon Railway Transit Camp for POW and Native Labourers' — was an improvised market where we could buy hard-boiled ducks eggs at five cents, mangoes seven cents, bananas two for one cent, pineapples 10 cents, lovely pomelos 10 cents, tobacco and cheroots almost given away, while for 40 cents you could get a four-egg- and-herb omelette with fried sweet potatoes and a cup of coffee, a substantial Mah Mee for 20 cents.

For the thirty-six hours we were there we lived like fighting cocks. Around the camp quite happily wondered muscovy ducks and drakes which reminded me of my Stalin.

To sleep that night we selected a level bit of earth, scratched a hole for our thighs, and made ourselves as comfortable as possible. I slept between Monty and Jacko. Sanitation in the camp had been nobody's business so that the area round about was foul, the stench was appalling and the place one mass of flies. No wonder dysentery raged.

The Thais would buy anything and I was able to sell quite a number of little things that I didn't want which gave me some badly needed ready cash.'

By early May 1943, rail had been laid up to Tarsao, 130 kilometres from the line's previous terminus at Nong Pladuk. However, travel by rail was not an option open to POWs ...

'The time passed very quickly here and it seemed that we had scarcely got out of the train before we were falling in on the road to start the first march. By Jap orders all marches had to be made at night owing to the number of POWs that had collapsed of sunstroke.

It was pathetic to see the huge loads of stuff the men

expected to carry. As well as their own belongings, every party of six men had to carry a six gallon container filled with our breakfast rice and fish; furthermore we were warned that those containers would be all that we should get for cooking purposes so they had to be watched as if they were made of precious metal.

At 8 pm on 16 May the great march started ...

The distance to be marched to the next transit camp was 17 miles. Every hour there was a halt of 10 minutes to allow stragglers to catch up. Beside the straggling column rode in rickshaws the Thai Chief of Police who filled his rickshaw with everything he could purchase off the men as they became exhausted and fell out to lighten their loads.

As time went on the column gradually straggled over six miles of road, but it was only guarded by six Japs, so certain were they that no one could escape or would dare to turn back.

With us we had three dogs from Changi; one fell out on the march having met a lady friend, one was destroyed by its owner, and the third one 'Nip' belonging to Jacko survived six months. A monkey that rode for the first hundred miles on its master's shoulder died at Tarsao.

About 2 am a man collapsed with cerebral malaria and Jacko stayed behind to look after him. The Jap guard came up to me and ordered me back to look after them both as he was afraid of the Thais leaning towards murder whenever they thought it would be to their advantage. We were, even at that hour of the morning, soon surrounded by children and inquisitive natives. As soon as I showed them my wound, their hatred for the Japs showed itself in thousands of little kindnesses. They brought us coffee, fruit and cigarettes. They produced blankets and a pillow for the sick man, and eventually, when we left, they forced $5 on me.

It was a lovely clear night, dry and cool. About 8 o'clock next morning a lorry came along which picked us all three up and dropped us at the transit camp.

This camp was located in an old school building compound near the river. A few tents with grass mats, capable of housing some hundred men was supplied for the 600, the rest having to find what shade and comfort they could under nearby trees. Breakfast consisted of a little rice and fish to which Jacko and I added our last tin of bully. Later we got a good meal both from the cookhouse and from native vendors. I had a large plate of fried caribou beef, three eggs, and a huge pile of fried vegetables, and a cup of coffee for 75 cents. We officers were housed in the old school building and were quite comfortable.

In the afternoon, after a comparatively good sleep, I went down to the river, a big one, full of teak rafts. Much to my joy elephants were at work shifting logs that had stuck. When moving a log near where I was wallowing in the shallows, the mahout made his elephant squirt me with water. The elephants were a grand sight. Their intelligence was quite uncanny. If they couldn't move a log after a careful poke or two with their tusks, they let out a trumpet when they were allowed to move on to the next one.

Seven o'clock in the evening of 17 May we fell in for the 25 miles march to the transit camp at Kanchanaburi.

Owing to the long marches through mud and tropical rain, when all one's possessions become a sodden heavy mass in a matter of minutes, the pulping remains of books — which had the feeling of a tin weight — were soon discarded.

Often when a man collapsed on a march I would go through his kit to see what he could discard. Sometimes he was carrying such things as weighty carved book ends which he hoped to give to his wife when he got home; often heavily made and carved photo frames, kits of tools and clothing far beyond his capacity to carry. My orders to leave and discard what was not absolutely necessary were always met with some remonstrances, but once the decision was made and the lightened man on his way again, I was always as relieved as the man. Up-country, books were like gold. Few people would carry anything except a very small volume as such things as mess tins and clothing had to come first and, the weaker we got, the greater the necessity to discard anything except absolute necessities.

Everyone was by now suffering from blistered feet which, combined with the pains and aches of beri-beri from which the majority suffered, made the march a hell. The transit camp wasn't reached till 11 o'clock next morning, everyone dead beat.'

The paper milling and market town Kanchanaburi thronged with bicycles, buffalo carts, monks, and vendors on foot or manning stalls either side of the main dirt road. Its 5000 residents mainly shacked up in wooden houses of the walled town, a legacy from forays by the Burmese attempting to sack Siam (the Burmese crossed the Mae Khlong river here at a place called Tha Makam). The Japanese significantly boosted the population with Kanchanaburi, the headquarters for the 9 Railway Regiment and Thai-side operations.

'Here, with a blazing sun, there was no shade except a few bushes and no shelter whatsoever except one tent for ten officers. The rest had to bivouac in the open. Fortunately there was a permanent cooking staff here that soon produced a hot meal for us; a really good stew, rice, and chlorinated water. I then slept till woken for the next meal in the evening.

Meantime at 6 o'clock an Australian party of 600 which had preceded us, set out on their 50-mile trek to Tarsao. The officers were then moved into the tents they had vacated.'

Kanchanaburi Transit Camp was typical of the conditions then prevalent on the railway — bad water, no shade, no shelter, with crude and almost non-existent sanitary conditions.

'As a sidelight on the period we were there a Dutchman had died of heatstroke and the Japs had ordered a Dutch officer to guard his body. This he had done by living with it in a barrel for 72 hours. When we arrived it was already partly decomposed. He was ordered to bury it eventually, but the Japs were quite indifferent. The Japs in Thailand had to go in parties of two or three for their self-protection as they were loathed by the Thais who had already accounted for many of them. So frightened were the Japs that they made POWs guard the camps as they were never molested, although they often had their gear stolen, even to haversacks and packs being cut off them while they slept.

Feeling better I dressed my foot, which had three times literally skinned, and hobbled round and found quite a number of 18 Division people who had been left behind as too ill to travel. They were later made to complete the whole distance of 200 miles to their camp.

Water for washing here had to be purchased at 50 cents a bucket. There was, however, a well worked by a bucket on a huge pole where, providing we could slip past the guard, we could get a good bath.

A good meal of the omelette *mah mee* variety with sweet coffee or boiling hot lime juice could be obtained from most of the native kampongs and were most acceptable. Prices here were about double those at Ban Pong.

After consultation with the Japs, an order was given that those who were not fit to march the next fifty miles through the jungle were to be left behind. On account of my raw foot, I was ordered to take charge of those remaining. There had, however, been so many casualties from previous parties during this march that, at the last moment, just as they were falling in, orders came through cancelling it. We were then told that everyone, including sick men, were to be taken to Tarsao by ballast train.'

For the Thai–Burma Railway, the Japanese ripped up the rails to Port Weld along with 150 miles of the east coast Malayan line, and sent them north. In all, 276 miles of running line and 57 miles of second line were stripped and shipped for use on the new Railway. The Japanese also appropriated one third of the pre-war stock of locomotives and half of Malaya's 5800 passenger wagons, and sent them to Thailand and Indochina.

'Next morning at 4 o'clock we had breakfast, were issued with our meal of rice and dried fish, and were marched to the station where we loaded on to trucks filled with gravel ballast.

At six o'clock we started and the journey over this bit of line — which was just opened — began. It was all through monotonous bamboo jungle with occasional large areas of very fine teak. The bird life was magnificent, the wonderful colours of jays, parrots and starlings the great masses of bright red and blue creeper that was growing over the bamboo made an indescribably beautiful sight.'

Interestingly, Pilk makes no mention of a 378-metre steel-arched bridge at Tamarkan just past Kanchanaburi, the only steel-and- concrete bridge in Thailand out of the 688 bridges constructed on the line by POWs. Made with materials largely purloined from Java, it was completed less than three weeks prior to Pilk's crossing. To the Japanese it was just Bridge #277, also referred to in one note from the Japanese commander to POW workers on it as 'Mekuron Bridge'. The world would, decades later, come to know it as the *Bridge on the River Kwai.*

Parallel to it was a wooden bridge completed in December 1942 for use by light diesel trucks. Swept away by raging floodwaters, it was operating again by January 1943.

These clearly became significant strategic targets for the Allies to take out. On 24 June 1945 three spans of the steel bridge were wiped out, thanks to the pioneering use of Azon bombs developed by the US Air Force, one of the first instances of guided bombs used in warfare. Twenty-three bridges along the line were destroyed in their three months of use. The Japanese repaired the breached wooden bridge, enabling them to continue transporting some of the 220, 000 tons of military supplies that were carried into Burma between December 1943 and August 1945.

'At one point a huge hamadryad [snake] almost tried to dispute the passage of the train. It must have been a full 15 feet long and for cold-blooded pride would take a lot of beating.

About 12 o'clock, having travelled some thirty miles, we were turned out of the trucks when they were immediately invaded by a horde of Chinese coolies who, under the savage beatings of their Jap masters, unloaded that train in record time. It was ghastly to see the fear that had been instilled into those peaceful people. [This was most likely Tha Kilen Station, based on the distances quoted.]

The train unloaded, we were ordered in again and proceeded another five miles to a siding where we got out and the ballast train went back. This was as far as a heavy train could at that time proceed [possibly Nong Pradai station].

The Jap guards told us we had two hours to wait, so I and Jacko wandered off to the nearby construction camp, where I found some of my own men who were working in the QM stores. They quickly produced a good meal for us, loaded us up with bananas and a tin of Jap M&B which they had looted, and filled our water bottles. These men were fit and comparatively happy but they gave me a list of my men that had already died. It was a long list and could by no means have contained all the names.

At 3 o'clock a little construction train came along; we were loaded in and proceeded the last 15 miles to Tarsao.

The railway here had been hacked through magnificent jungle and mountain scenery, and at times it was carried on ledges overhanging the river below [Wang Po viaducts]. Several engines and trucks visible half under the water below told of the dangerous nature of the track. We were glad when once more we were on comparatively level land.

We reached Tarsao, the name given to the town or construction camp at the then railhead, about 4:30 pm, and were made to sit by the railway track while our head Jap went off for instructions.'

This might actually have been Wang Yai station, about five kilometres south of Tarsao station. It was a shorter march to Tarsao work camp from here.

'We arrived at Tarsao in the evening of 20 May. All day the heat had been unduly oppressive with banks of black clouds on the horizon. At six o'clock the first storm of the monsoon broke and, within a few minutes, everyone that couldn't find cover was soaked.'

That monsoon unleashed its spiteful rain for 140 consecutive days.

'I fortunately found shelter in a signalman's hut but a number of men who sheltered under an old railway tarpaulin truck cover were chased out into the open by a Jap with a bayonet. Shortly after the storm had broken we had orders to move to the transit camp some one mile distant. I stayed behind to see our dysentery cases (three) into hospital. Before I could get them admitted it took me an hour to find the hospital and, when I did, it was only by a Jap officer intervening that the hospital authorities would consent to taking them. There was an order, very strictly enforced, that no one in transit was to be admitted to any of the hospitals en route. The Jap officer I had met, however, signed the admittance certificate, for which I shall always remember him. He was one of the few that ever went out of his way to help a POW.

After all the trouble I had taken to get these men proper medical attention, they discharged themselves from hospital next morning and insisted on continuing the journey with us. Within a fortnight they were all dead.'

At Tarsao—designated hospital camp for D Force—a graveyard called St George's was established over the road, east of the camp. Subsequently, as the toll rose in what was widely considered as one of the foulest camps on the whole railway, St

Luke's was established near the river south of the camp, a third graveyard slightly inland of that, and a fourth one beside the river just north of St Luke's. In all, a thousand POWs would be buried here.

'Eventually, in pitch darkness, soaked to the skin I found the transit camp which once again proved to be only a few tents in which the sick had been herded; the rest were sitting miserably under the trees, trying to shelter from the deluge under their groundsheets, that is those that had them. No meal was served that night. The Jap guard commander came across Jacko, Monty and myself and took us into the guard tent where he fed us on his own rations and cleared a corner for us to sleep. I fell asleep almost at once and woke up in the morning to find that I had been lying in a puddle of water some three inches deep.

Next morning we were told that no meal would be available till 12 o'clock so I went off to forage for myself. I eventually ran across some of my 6 Norfolks men who gave me some fruit and biscuits. They then told me that Padre Alcock and Colonel Knights of 4 Norfolks were in a hut just down the road. The reception they gave me was almost overwhelming in their kindness. While the Padre gave me a razor, towel and soap, and a bath, Colonel Knights produced hard boiled eggs, biscuits and mugs of hot coffee. Never have I enjoyed a meal like that!

He was surprised and furious that I had been allowed, even at my own wish, to come up. The work those two did for their men will be one of the great memorials of the POW existence. For myself I shall never forget their help and kindness.

Fed, shaved and washed, with my clothes dried, I went back to our camp where later we had a poor meal of thin pumpkin stew and rice. Afterwards I went down to the river, about half a mile away, with Carlton. We bought eggs off a barge, had a bathe, Carlton washed my towel and spare shirt and — while they dried — we slept on the grass.

The sun fortunately came out and dried out the men. That evening we had a meal of reasonably good stew and rice and half a pint of tea. Our water bottles were filled up. At 6:30 the sick who were considered unfit to travel were paraded before the guard tent for examination by a Jap doctor. The doctor, when he appeared, was a Jap private who, without looking at anyone, started beating them with a bamboo. In the end he shipped the whole lot back into our column waiting to move off. No one in transit, even though dying, was allowed to halt.

Our march that evening of 22 May started in a fine drizzle. By nine o'clock the full force of the monsoon had broken. The roads, nothing but mud tracks, were quickly ankle deep in water; as the storm continued the surface changed to mud of such a gluey consistency that men pulling out their feet often had the soles of their boots sucked off. We struggled along through the deluge all night being guided by the flashes of lightning, while the roar of the thunder never ceased, which, with the hissing of the rain, meant that we had to shout to make a man four feet away hear.

About four o'clock in the morning the rain eased to a drizzle and by dawn had ceased. Many people fell out and slept by the track- side to continue the journey next day. Others jettisoned some of their most precious belongings such as blankets, which, sodden with rain, had become unbearably heavy. Many a time I pulled a man to his feet only to see him collapse a few minutes later. I saved Pat Mallins, an old Salopian School fellow, from drowning in the mud. He had fallen face down with his pack on top of him and literally could not get up.

In spite of the rain, the loss often of their only boots, soaked to the skin, with festering sore feet, the men never lost heart. They cussed and swore, they laughed, joked and sang. It was a magnificent display of pluck and far more heroic than anything I saw in battle.

One rather amazing case occurred. An officer and six men unable to go any further made a fire and lay down to sleep for a few hours. The only covering they could find was a Union Jack one of them was carrying. This they spread over themselves. Early next morning a Jap officer passing by came over to see what it was. When he saw six men apparently dead lying under the flag he turned pale and yelled 'Cholera' and, with that, he and his brave men fled down the road and were last seen running hard.

The last night's march was about twenty miles, over a precipitous mud track which a week ago had been eighteen inches deep in light dust but from now onwards the monsoons were to turn it into such a mud bath that no transport was able to proceed along it and all those walking had to keep to the jungle.

At dawn about 7 o'clock next morning, we, that is those of us that were still on our feet, arrived at a bamboo-stockaded and tented camp. Here we were halted and told that we had reached our destination.

I was standing completely exhausted, looking for something comparatively dry to sit down on when a bearded, lean and dirty figure in torn and muddy clothes came out of a tent and said, "Hello, Pilk."

"Your voice is familiar but I cannot recognise you," I said. It was Graham Henderson of Sungei Ujong Estate [a fellow planter from Seremban, Malaya]. It was only by their voices that we often recognised the scarecrows that were our friends.

He took me into his tent, filled me with hot coffee made from burnt rice, fed me with rice porridge, and finally gave me what was worth its weight in gold, a fill of tobacco, all mine being soaked through.

Shortly afterwards I watched him and six hundred walking, dirty, unkempt skeletons herded off to work.

Later Monty turned up. We were all served with a mug of hot water and informed by a Jap officer that we had reached our destination.

Of the 600 that had started out only 200 had arrived by 8 o'clock, the rest came in dribs and drabs during the morning and even three days later.

So ended the second part of what is now termed our Death March. I arrived at the end of our marches with the barest necessities of life: a mess tin, my watch, and a hairbrush!'

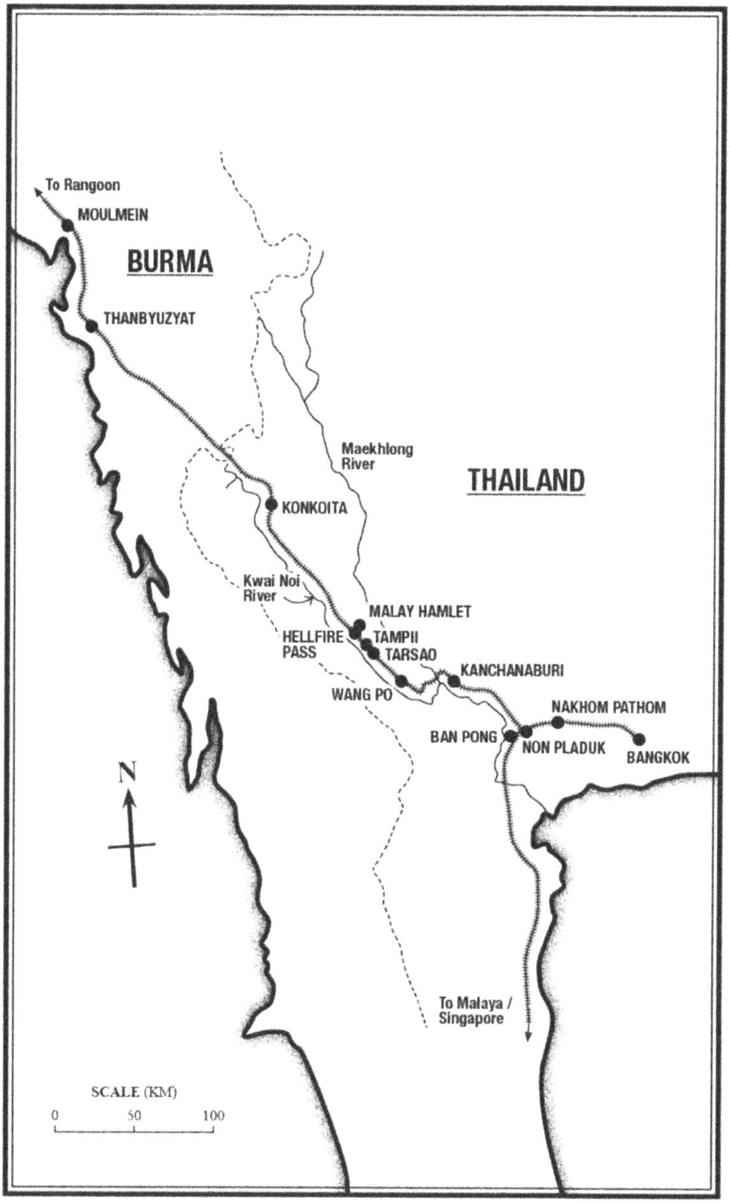
To Rangoon
MOULMEIN
BURMA
THANBYUZYAT
Maekhlong River
THAILAND
KONKOITA
Kwai Noi River
MALAY HAMLET
HELLFIRE PASS
TAMPII
TARSAO
KANCHANABURI
WANG PO
NAKHOM PATHOM
BAN PONG
NON PLADUK
BANGKOK
N
To Malaya / Singapore
SCALE (KM)
0
50
100

The railhead station at Ban Pong where they disembarked from their cramped cattle carts. (AWM P00761_029).

The area around Tampii was delightfully flat, but they could see the challenge rising in the ranges to the north, where they'd be working next. (AWM P01910_022)

12 DYING A SLOW DEATH

Jungle Camp Nakamura Thai/ Tampii, Thailand, 22 May–26 June 1943

'The place where we had now arrived and where we were to live and cook for the next six weeks was an area of giant bamboo jungle with a good stream running through it. Nearby and surrounding it were huge expanses of the most glorious teak forests. I think that next to an oak or ash, a teak tree is one of the most beautiful. Tall and straight with the reddest bark and large soft light green leaves; very soft to the touch.

The district itself was on a great ledge some two miles deep with behind us huge limestone cliffs and hills, very rough and jagged against the horizon, while in front of us was an almost sheer drop of 1000 feet [actually about 600 feet] to the river below which we could see winding its way through dense jungle. In the distance some 50 miles away were the Burma Hills, a magnificent range of purple mountains. We were, I believe, here on the crest of the dividing hills between Burma and Thailand and about 2500 feet above sea level.'

This piece of countryside is deceptive: it presents as high-country but in actuality rises from just above sea level to around 1500 metres in the highest mountains.

'Morning after morning I was to see the mists gradually clear from the river valley and the sun come out. A lovely sight of which I never tired.'

The railway itself was being constructed on a climbing gradient along the edge of this ledge. Initially work on the line was incredibly quick and roughly on schedule as the land was moderately even. Then came challenges of rugged terrain such as the cliffs around Tampii (meaning 'Ghost Cave'). Going up and over was not an option.

'Very exhausted on our arrival we lay down and slept for a couple of days when I was woken by Dr Young who had found a tiny Thai shop which was selling lumps of *gula malacca* on leaves for about 50 cents a pound. We all ate a pound and felt much better. I then wandered off and had a wash and shave in the stream and cooled my feet in the cold water.

When I got back I found that Monty had received instructions from the Jap officer to start constructing our camp at once. We were led to a piece of virgin jungle and told to clear it of all bamboo small trees and undergrowth. Crude, half broken tools were issued for the purpose. At the end of two hours the site was cleared and all leaves swept up, leaving nothing but bare earth. We were now told that there would be no meal for us till the cookhouse was built so all hands were put to this job. Bamboo frames were cut and fitted together, a matting roof was fixed to the eaves and four huge iron pots called *kawalis*, each capable of taking some twenty gallons of water, fitted into circular raised mud walls. The fires were lit and our cooks started preparing our evening meal.

In the meantime some dozen tents had been dumped. These were in nearly all cases old British Army tents, full of tears and rotten to the touch. In no case was a tent complete. There were either the inside walls or the outside flies. These were quickly erected.

Next under instructions from our guards, bamboo platforms were constructed inside these tents about eight inches off the ground and planked with thin bamboo split into slats. These were to be our beds.

Everyone was by now dead beat, with darkness drawing on. The Jap officer came to inspect the work, appeared thoroughly dissatisfied, and beat up anyone handy. He then said that a party of 200 men were to be ready to start work on the railway next day but, finally, when he was convinced that there wasn't a man fit to go, he said two days would be granted to make the camp habitable.'

Prison camps took their orders from the Railway Corps. However, because precedence was given to the over-arching construction orders issued by Imperial Headquarters, prison camps often sent sick prisoners out to work.

'Our meal of plain rice which I was able to supplement with a hard boiled egg was then served, but we were all so ravenous by then that even rice was good.

No sooner was the meal completed than the monsoon rains

started again turning the whole camp into a quagmire. We were, however, by now quite indifferent and turned in to sleep heavily till eight o'clock next morning when we were woken to another meal of plain rice and a cup of hot water.

The Jap guards had every man working, however sick or ill he was. I had four men who could scarcely stand in the jungle. On this day the cookhouse was completed, the stream diverted to run past our camp and latrines dug. We were given two small meals of plain rice.

Next day, our last day before work was to begin, was spent in constructing drains, improving the cookhouse and erecting two brand new Jap tents, one for the guards and one for the officers. At the same time I took six men with axes and felled all the big trees round so as to let in light and air, and almost immediately the sun brightened from tropical gloom to a cheerful open jungle outlook and I am sure was a great factor in keeping our death rate down in this camp.

By now my feet and ankles were very badly swollen with beri-beri and I could scarcely walk so I was told to slip off into the jungle and rest. I slept a full six hours till the evening meal. Our rations from today were supplemented by a queer type of dried or dehydrated broccoli that swelled when boiled to about ten times its volume. It looked just like shredded pieces of our gas capes and even became known as 'gas cape' stew. A really filthy concoction, quite tasteless.'

Thai broccoli (*pak kana*) is in the same family as regular broccoli, Brassica, but instead of a thick stem with huge flower heads it is a big- leafed green with flower buds that resemble regular broccoli.

'Next day work began. As the day was typical of any other I will give it in detail. At 4 am reveille was sounded. The Jap guards never deigned to keep awake at night so our cooks had to act as night watchmen and wake both us and them. One usually woke to the sound of a piercing whistle while, in between blasts, the blower would yell "Wakie! Wakie!" as he went from tent to tent shaking the occupants in to consciousness. The monsoons were now in full blast and all night long we could sleep to the thunder of rain, but usually it cleared by the time we had woken, to a fine drizzle. At 4:30 the men lined up for their breakfast. A pint of rice porridge with a teaspoon of sugar water. At 5:15 everyone drew their midday meal from the cookhouse, which consisted of a pint of plain rice and a piece of dried salt fish about one inch square. At 5:30 big water bottles made of lengths of giant bamboo, each water bottle in three sections with the intervening pieces knocked out, had to be filled. These were remarkably heavy and contained a surprisingly little amount of water. At 6 o'clock everyone fell in on the so-called parade ground by companies. We were divided into four working companies of which I took charge of one. The ground was ankle deep in slippery mud and it would still be drizzling. The guards having checked our numbers, when we had to number in Jap, checked each man to see that he had his water bamboo filled. About 6:30 the Jap officer, Oswald, who was in charge of this section of the railway would appear and detail us to take the men out to work. We were never accompanied by guards, so impossible was escape considered.

To get out of camp and to the working site entailed a dreary trudge through the drizzle of about two miles, along roads always ankle deep in mud but just as often knee-deep. We solved this – after most people had lost a shoe, a boot or had the sole dragged off – by going out barefoot for the first mile till we reached the rocky path leading to the railway and here we would put on our boots. Arrived at the railway cutting about 7:45 the men would hang up their ground sheets and food and would then fall in with their bamboo water bottles. At eight o'clock the Jap RE Sergeant appeared, checked the numbers, ordered the water bottles to be put near the petrol compressor which leaked badly and had to have a constant supply of fresh water, and would then take the men down to the railway cutting below.

All the work on the railway done by the POWs consisted entirely of blasting and hacking and cutting through the limestone cliffs. To do this half the men would be kept at work with picks, shovels and earth baskets removing the soil, while the other half worked in pairs drilling the rock for cordite charges. This they did with cold chisels about 24 inches long, one man holding the chisel while the other wielded a sledgehammer. In order to clear the hole of the pulverised rock, some of the precious water brought out in the bamboos was used to wash it out. This operation was known as the 'hammer and tap' and was preferred, although harder than the 'pick and shovel', as it was on the task each man being given so many metres to finish when he could go home.

Work on the line would go on without a break till 10:30 when half an hour was allowed while the holes were filled with cordite and blown. As the rock displaced by the explosions rumbled to earth, the men were again herded back to work till 12:30 when an hour and a half was allowed for the midday meal.

By now I would have a big fire burning with a container of boiling water. I would give each man his share, about 1⁄4 of a pint, after which I would hand them back their little lots of fish which I had grilled for them while they waited. In about five minutes their scanty meal would be finished, a cigarette rolled out of local tobacco and the men would talk till 1 pm. I then made them all lie down and sleep till 2 pm when I had to return the full number of men to the top guard. The afternoon was a repetition of the morning, a break of 15 minutes being allowed between 4-4:30 pm. If all went well work stopped at 6:30, the men fallen in and checked when a tired and weary line of men made their way back to camp, which they would reach just as night was setting in. After a hurried wash in the bathing pool they would draw their dinner of one pint of hot water, a pint and a half of rice and a pint of gas cape stew, later thickened with dehydrated potatoes.

At 8 o'clock, in almost total darkness, the men fell in for the last check of the day, after which they would settle down in their leaky tents to pass the night away in cold misery and discomfort, the rain thundering down on their tattered tents, and the ground outside — which had dried out during the daytime — a sticky quagmire.

The days passed in this fashion without ever a rest. It always rained at night. It was generally fine during the day. If it wasn't, that was just too bad. The work went on wet or fine.

As regards the Jap administrative staff, as soon as we had arrived at this camp we came under their engineers, while the actual running of the camp was continued by our Korean guards, acting under instructions from the Jap POW HQ. This meant that, providing our guards supplied the requisite number of men, they were left alone to enjoy themselves. If they didn't, they got beaten up by the engineers.'

Having forcibly annexed Korea in 1910, the Japanese military drafted 240, 000 Koreans to serve in the military and in subordinate capacities as civilians. Some 3000 were dispatched, after basic training, to serve as prison guards throughout South-East Asia. Some of these guards were formerly soldiers, others reportedly released prisoners. On the totem pole of the Imperial Japanese Army they were automatically below all Japanese of any rank, and viewed only slightly above the POWs.

'With one or two exceptions our Korean guards treated us, according to the Jap point of view, very kindly and did all they could to help us. There were however, sadistic ones amongst them that always walked about with bamboos, beating up the men for the least fault, actual or imaginary. Later our Jap sergeant weeded them out and to his eternal credit beat them up.

Our Korean quartermaster was always on the make and would invariably try to give us short rations, helping himself and his companions to as much as they wanted. In this way they stole nearly all our sugar ration, all our salt ration and a great proportion of our dried fish. What made their behaviour more disgusting was that they could never eat all they had stolen but would throw most of it away.

Generally the guard would get hold of some bottles of Jap or Chinese brandy which, acting like an incentive on them when they became drunk, would turn them into mad beasts and — unless one wanted a savage beating up — one had to keep well away. Not so easy when their tent was only separated by a few yards from the officers.

The Korean, under whom I administered my company, was most considerate. I never knew him to beat up a single man however aggravating he had been and to me he always showed the greatest respect. Long after I had left the camp he trudged many miles through the mud to see me and make sure that the rumour that he had heard that I was dead was not true.

The only real incident at the camp with the Koreans was when Monty had a fight with one of the guards; not a very sensible thing to do, as he got badly beaten up. The rest of the guards who disliked the Korean who had caused the trouble, however, quite approved and when honour was satisfied all settled down amiably again.

As mentioned, this area was under the control of a Japanese officer called Lieutenant Nakamura and, as he figures largely in the atrocity reports, I will try and give a picture of him. He was small, thin, with big eyes hidden by huge spectacles and from his queer way of walking was known up and down the line as 'Oswald' after the rabbit. He was one of those half-educated people with great ambitions in life that had been frustrated. What at one time must have been a very kind disposition had been turned to moody melancholy by possibly many reverses and slow promotion. He had only recently been promoted from the ranks and was obviously trying to establish a reputation for himself. On occasions he could be the soul of kindness, sending us in loads of blankets, boots, Dutch clothing, changing our money (and even lending us some), seeing that we got a cigarette ration, allowing parties to get eggs and he could even be gentle with the sick.

At other times he was, if the slightest thing went wrong, a raving lunatic. At first he insisted on every sick man, even if he were dying, being paraded at 5:30 am. Anyone with a visible sickness such as beri-beri or cuts he automatically excused work, but hidden diseases such as dysentery and malaria got no sympathy and, with blows and kicks, they were sent out under an especial guard to do an extra hard day's work. To me he generally showed the greatest consideration because of my shoulder wound, and whenever I had beri-beri he refused to let me go to work. This was more often than not. On one occasion he personally helped me home when my feet were so swollen that I could hardly walk. Another time he beat up two of my men who he thought ought to be helping me.

Against this must be set the opposite treatment. Once on the early morning parade he muddled up the gangs and sent them to the wrong working sites. His fear of 'losing face' overcame him and he went down a whole row of men savagely slapping them over the face, finishing up with me. Later he offered me a packet of cigarettes to show that he was sorry. I didn't dare refuse as on three occasions when I had done so I had been beaten up by the would-be donor for 'arrogance'.

On 8 June I was suffering from a very bad and acute attack of MT (Cerebral Malaria) and when the early morning sick parade fell in I was only semi-conscious and was helped out by other officers. Oswald on this occasion singled me out as a *shoko* (officer) and evidently kicked and beat me with ungovernable sadistic fury. I, being unconscious, knew nothing of the matter till later I was being helped back to bed by Dr Young with orders not to move till I was better. Later Monty, in a furious temper, told me what had happened. This incident is one of many that figure in the atrocity reports.

His favourite method of dealing with dysentery cases on his bad days was to kick them in the stomach. I never knew him to send anyone out with sores or swollen legs, but he always insisted on all bandages and dressings being removed. The last I saw of Oswald was when he had a fight with one of our guards on the day we left this camp. It appeared that he had expected us to have left the night before, but our Koreans had thought otherwise. After a few words he tried some jiu-jitsu on this guard. He was very proud of his skill in this art but he had met his match. Our Korean was still more expert and threw him, to our delight, time after time in a clump of thorny bamboos, which must have been very painful. Foiled, he left in a raging temper.

I was, however, very lucky in regard to the *guntsa* (sergeant) who administered my working gangs. Known to us as Beardy he was a real gentleman among that sadistic crowd. My party worked alongside a party of Malay Volunteers from the adjoining camp and, when going out to work the first day, I was fortunate in finding a number of old friends amongst them who told me how to create a good first impression. They said if I would do this I and my men would be shown every consideration.

Arrived at the site, therefore, I fell in with my men and when the sergeant arrived made them number in Jap. I then repeated to him with a smart salute the last number. At the end of the day's work I repeated the performance. He was delighted and, pointing at me said, "You number one verrah good-ah." Beardy only on very rare occasions beat up a man and then he nearly always 'had asked for it'. Before doing so he always brought the culprit to me, explained his fault, and made me watch the punishment which was never savage. He never allowed his underlings to touch a man. He never allowed me down on the line but told me to keep a fire and hot water going and, to me, he would send men that became sick, to look after. No other working parties that I ever heard of were allowed this privilege.

Invariably, if I asked permission to send a sick man back to camp it was granted. As the men got weaker and the gangs dwindled to a mere nothing, he showed his understanding by letting the men work in their own time, never hurrying them, providing they were doing their best. In order to try and keep the men fit he allowed us to build bamboo and mat shelters for them. He was the only ganger who allowed work to stop while it rained.

Every day he would shyly present me with a few quinine tablets and some dysentery pills for the men, which must have come out of his private store.

During the lunch hour he and his subordinates, encouraged by his example, would send us over some of their lunch which helped us considerably. The only work he ever asked me to do was to go back to the camp occasionally to get more petrol for the compressor motor. His cigarettes he invariably shared with us. He was a gentleman amongst Japs. If there had been more of his kind the death rate would not have been so appalling.

This treatment must be compared with that of other officers with their tales of men savagely beaten, made to stand for hours with a crowbar in either hand over their head; made to run at the double with tip trucks on bare feet over stony tracks; cut, bruised and maimed by blows from shovels and picks. Gangs kept out at work for 24 or even 36 hours at a stretch without a pause and then only allowed four hours' stop before they marched off again.

Our food was nothing but rice, gas cape stews, dehydrated potatoes (a luxury), and a little sugar. Every now and again a little oil, but this was generally stolen by the Japs for use in their lamps, and dried fish. At long intervals our administrative HQ would succeed in buying us a pig. Invariably the Japanese stole half for ten guards. The other half we were allowed to retain for 600 men. Oswald did, however, allow us to trade with a Chink who had a tiny stall some two miles down the road and who, by means known only to the Chinese, succeeded in getting in supplies of eggs, *gula malacca*, peanut toffee, and a sweet native cake called *gulo* cakes.

Every penny we could raise was used to purchase these supplies, the men being allowed their share on credit. As I also had an outside source of supplies through the Volunteers, I was still better off, and, as I had been given two pots of Marmite to use in case of necessity by the eye specialist, I was able to keep the wolf at bay.

The diet here was not so much starvation but slow death through lack of vitamins.

The cookhouse was run by Lieutenant Bowman who did a noble job of work, particularly as regards his methods of continually stealing masses of fish and vegetables from the Jap supplies, which went a long way to keeping the death rate down. When one saw the crude conditions under which he and his staff had to work, one can but give him credit for a job well done.

Sickness we had with us from the first day. Our first dysentery case died three days after we reached the camp. At that time there was no hospital tent and permission to send him back for proper treatment to Tarsao was refused. He died, therefore, in a little lean- to shed hastily constructed out in the open. As dysentery increased, even Oswald could not ignore it and we were supplied with an extra tent which was erected on its own outside the camp. Soon this tent had to be augmented by another. For dysentery there was no cure as we had no Epsom salts and it was impossible to diet them. Small supplies of milk did come at rare intervals but the quantity was like a drop in the ocean.

Sixty or 70 motions a day were recorded in the macabre dysentery sweepstakes run in various camps.

As malaria began to make itself felt, the sick rate rose alarmingly. Our quinine supplies were rapidly exhausted while supplies from the Japs were inadequate and of very poor quality. No man could ever have a full course, so no man was ever free from malaria. Before we left, the malaria rate was 100 percent.

Either a man fought for his life on his own or he gave up and died.

Tropical ulcers, due to weakening diet and malaria, soon made their appearance but, thanks to a little M&B 693 paste [sulphanilamide, an early wonder drug] that we had, were generally curable. As dysentery took a more firm hold on the camp the need for strict sanitation became more essential. As the men got weaker they began to cease to trouble to try and get to the latrines, with the result that flies increased alarmingly. Eventually I caught a man who had not bothered to go to the latrine and — taking jungle law into my own hands — I knocked him unconscious with a blow on the jaw. The man has lived to tell the tale but I don't think he will. After this there were no more incidents of this kind and, once more, the men struggled gallantly to keep the camp clean.

Early in June cholera was reported both above and below us.

After this, the men were made to wash in boiled water, all eating and cooking utensils were sterilised before meals in boiling water, and the men lectured continuously on sanitation. The camp only had three deaths from cholera and only 26 deaths in all. A record for the line. To this all credit is due to our three MOs — Dr Jackson, Dr Young and Dr Moss — and the RAMC orderlies who worked night and day, often very ill themselves.

The dead had to be buried by parties of those least sick and such officers that were available. By Jap order they had to be stripped of their clothing, which was boiled and then given to those men who needed it the most. I think one of the saddest tasks I had was reading the shortened Burial Service. Their deaths were so tragic and unnecessary.

As regards myself and my 21 other officer companions, we were crammed into a tent made to hold six. Our bed and living space was 16 inches by six feet. It was impossible for everyone to sleep on their backs at once. Few of us had blankets to begin with so that we had, to keep warm, to wear all our clothes. Fortunately I had my waterproof cape which served me well. Later I was fortunate in drawing one of the Jap blankets that they sent in.

In our food there was no discrimination allowed with the men, the only difference being that our share was served to us separately and was divided out among ourselves by ourselves. We had, however, many little advantages. For instance, we had a small cleared space that we had made where we always had water boiling for coffee and washing purposes. By each subscribing a sum of money we were able to supplement our diet with extra eggs, sweet coffee, etc. We were also allotted a few tins of tinned vegetables and jam. We had no batmen, the officers not out at work during the day doing the necessary chores.

In spite of the acute discomfort of this period there was a very real spirit of cooperation and goodwill amongst us that makes this period a pleasant one in our memories. It should be remembered that there was as much illness amongst the officers as the men, but rather than let our companions go to the hospital tents, we nursed each other. Monty Smyth kept a very good balance amongst the officers and his self-sacrificing work was an incentive to everyone. By Jap orders only two doctors, OC Camp (Monty), officer i/c cookhouse (Lieutenant Bowman), and officer i/c working parties (Pat Mullins), were officially allowed to remain in the camp during working hours. But there were always three or four sick officers as well.

Every day down the squelching muddy road running past our camp went parties of POWs going to construct the railway higher up. Dirty, often bootless, their clothes in rags, unshaven, footsore and covered with bandages, with all they possessed in the world – and which might mean life or death to them – strapped to their backs, they went up the line singing and joking. One pleasant sight was a Dachshund following his master whom he never allowed out of his sight.

On other days long lines of weary Chinese and Tamil coolies went by. Sick and ill, they were allowed no rest. A more pleasant sight were Jap soldiers making their way to Burma, dispirited and covered with mud, dragging their little commissariat carts behind them. Often the carts got so bogged in the mud that they had to be lifted bodily over by willing men. At other times these carts were drawn by nice little ponies and oxen who, when they got stuck in the mud, had to be carried to firmer land by their masters. Sometimes parties of cavalry went by on well-kept ponies. Always the Jap soldiers were loaded with huge bundles of loot that they would rather die than abandon. The contrast between the victorious, disillusioned, dispirited, well- fed Japanese and those apparently happy and unbeaten POWs was always a sight that made us feel proud.

The cool jungle shade in which we worked and the ducks eggs on which I spent every penny I could lay hands on soon had a beneficial effect on my eyes. Against this, however, I suffered badly from beri-beri and aching feet which incapacitated me for a lot of the time. I also developed a few tropical ulcers on my wrists and ankles which, thanks to prompt treatment with M&B, cleared up in record time. I didn't contract dysentery but I did have two very violent attacks of cerebral malaria, which finally Jacko brought under control with some of a precious hoard of Atebrin.

By June 10 the men fit for work had shrunk to a mere dozen or so. All the rest were in hospital. So bad were conditions that the Jap authorities withdrew us from work on the line, turning our camp into a hospital camp only. Those men that could still work were used to carry up the rations from the river six miles away.

Through all this time news bulletins were brought to us, sometimes on slips of paper, but, generally, by word of mouth, so that we always knew about the outside world.'

It is estimated that dozens of clandestine shortwave receivers were in operation by the POWs up and down the Railway (six in Tarsao camp alone). One such snippet they might have heard on 18 June was that Field Marshal Archibald Wavell had become the Viceroy of India, despite a poor showing as Commander-in-Chief in India and Burma (which resulted in any case in his promotion to Field Marshal in January 1943).

'By 20 June Cholera had attacked the Australian camp six miles up the line to such an extent that they could no longer supply the engineers with their daily quota. As our men gradually got better, the Japs, therefore, transferred them to this camp. Every few days small parties would be marched off into the blue to face an ordeal that can only be described as ghastly. With one of these parties went Carlton who, in spite of my getting the Japs to let him remain behind with me, said that his place was with his companions. At this time I was suffering from a mild attack of Phlebitis and found walking difficult, but by 26 June I was fit enough to accompany Monty and a large party of our convalescent men to the new camp.'

Although it does in no way exonerate the Japs, it should be borne in mind that in most cases they were carrying out orders from their superiors who, in their turn received their orders direct from Tokyo, who neither knew nor cared under what conditions the railway was being constructed.

By the end of June 1943, tracks had been laid up to Tonchan station at the 139 kilometre mark. That same month, US convoys from Pearl Harbor exceeded twenty in a month for the first time. An intense neck-and-neck tussle was developing between the advancing Americans and the supply routes to the Japanese offensive effort in the Pacific. Construction work on the Railway was in full swing: 61, 000 Allied POWs and over 150, 000 romushu Asian labourers had been co- opted into the enemy's war effort. Around 45 percent of all prisoners of the Japanese were condemned to working on this notorious project as 'white coolies'.

All that mattered to the Japs was that the railway must be completed by August in anticipation of an autumn offensive in Burma.

The increasingly mountainous terrain towards Burma required cuttings like Hellfire Pass, or bridges like the Hintok trestle bridge to be constructed. (AWM 157859) (AWM 122309)

13 A PROGRAM OF DELIBERATE MURDER

Jungle Camp Malay Hamlet/K2, Thailand, 26 June–14 September 1943

'Ten o'clock found a party of some 120 men, Monty, myself and three officers, filing out of the camp, the men laden not only with their own gear but – slung between them on bamboo poles – our dixies and tents. The night before, it had rained and all our canvas had been soaked but the Japs insisted on striking the tents still wet, and everyone knows what a weight wet canvas is.

The roads to the next camp were quite impassable for mud and the only navigable route was along little paths two feet wide made by other parties along the edge of the road in the jungle. In places the trees were so close together that the tents, which take up a lot of room and are bulky objects, could only just be squeezed through. At other places it was possible to walk along the railway cutting, here being constructed by Tamils, but the last two miles had to be done along the road; there was no other route.

On the way up in a damp little camp I found Colonel Houston, the surgical specialist, and Major Agnew of 18 Division HQ, with about 40 other men living on rice and dried fish. Taken ill on their journey through to Burma, and too ill for even the Japs to drive further, they had been dumped here.

Recovered from malaria and sore feet, they were now waiting to proceed onwards with the next party. In the meantime their Jap guards allowed them no facilities to purchase extra food, available at the riverside some two miles distant. It was delightful to meet them again and I rested here for about an hour, hearing all their news, drinking coffee and listening to reports of their parties.

Before I eventually arrived at our new camp I was herding, with jokes and threats, three men who kept collapsing by the roadside, who, although I loaded their little kit on to the backs of other already overloaded people, could still only just walk with help.

Three days later two of these men were dead with cholera, which they had already contracted; the other lingered on with dysentery for some five or six weeks. He too had died after making a very gallant fight for his life.

Through mud up to their knees, often up to their thighs, overloaded with sodden tents, the men struggled to their new camp. A journey that should have been done easily in two hours took five or six, and even then the last of the stragglers did not arrive till six o'clock. Eight hours for six miles.

Our new camp called Malay Hamlet, but more commonly Kanyu Construction Camp No. 2, was already in being, our previous parties under Captain Shakespeare having already done the clearing of the bamboo thickets. All that was necessary, therefore, on arrival was to make men, collapsing with fatigue, pitch their tents, cut bamboos to make their bed platforms, and to cover these with thin bamboo slats. The work that should have taken perhaps an hour with fit men, took three hours and even then only half finished.

The officers' tent, a Jap one, in reasonably good condition, was already standing complete with bed *chungs*, as they were called, so that all I had to do was to grab a space and dump my earthly possessions on it.

When I arrived at Malay Hamlet I was told that Carlton had contracted diphtheria. I found him in a little lean-to hut that had a stretcher for his bed. The serum that would have saved his life was not available and the Japanese refused even to attempt to get some. He was, however, making a good fight. Whenever supplies were available I got him eggs and milk which I made him swallow under my eyes. About 10 July he could still swallow but vomited up everything he got down. I tried eggs whipped in tinned milk which he was generally successful in retaining. All the time he was very cheerful and could talk with ease. He was over the worst and there was every hope of his recovery. I last saw him on the evening of 10 July. On 11 July I went down with a cracking dose of malaria. On the night of 13 July there was a terrible thunderstorm and some time during the night he died of diphtheritic heart failure. He was buried next day with Monty and I as pallbearers. Colonel Oakes read the funeral service. Amongst his belongings was a watch his wife had given him for a wedding present and a small diary. Nearly all the last entries referred to me and his delight that we had stuck together. These, all that were left of his worldly possessions, should reach his wife through the Red Cross if the Japs have kept their promise. I was not allowed to retain them.

So passed in miserable and degrading circumstances the most faithful and loyal servant and friend man ever had.

The site of the camp was on a beautiful sloping hill with a stream running and gurgling some twenty yards away. It was surrounded by dense bamboo thickets while behind it towered sheer and majestic, huge rugged, jungle-clad limestone cliffs. A sight that even when things were at their worst always pleased me.

Across the road, over a foot deep in mud, on a similar hill was the Australian camp. The air when we arrived was thick with the nauseating smell of burning bodies, the day's batch of cholera victims were being cremated.'

By the time Pilk arrived at Malay Hamlet — also confusingly known as K2 but not the same camp as Kanyu 2 which was further south — there had been 31 deaths at this camp. The first few cases were Asians, then Allies started to die in large numbers.

'Matters were indeed grim, the Australians were dying five or six a day. Cholera had already spread to our camp and still the engineers demanded their quota of men; if it couldn't be filled with fit men, sick men must go, but the correct number must go out and the correct number return, even if this meant producing a dead body.

The crime that the Japanese authorities deliberately committed of sending comparatively fit men into a cholera camp without allowing them the ten days necessary to be inoculated, when the serum was already there, is one of the blackest on their records of the railway. It cost the lives of nearly 50 percent of our British force of three hundred.

There can be no doubt that if the Japs had supplied the cholera vaccine immediately the first case was reported, and which they had ready, the death rate would have been comparatively light. As it was, most of us who survived owe our lives to these inoculations. The MOs I believe are of the opinion that at one time everyone in the camp had cholera in varying degrees of mildness.

The cheerfulness of the cholera and dysentery tents was always unbelievable. These torn, ragged and leaking tents — erected in a little bamboo clearing with tall and gloomy jungle trees shutting out the bright sun — were often so crowded that there was little bed space. The beds were nothing but bamboo slats raised on bamboo poles on a sort of platform a few inches above the sodden ground. The iller the man, the more cheerful he was. Their gratitude for the little one could do for them was payment indeed for the hard treks to the river canteen through six miles of mud, jungle, and precipitous paths to buy them eggs, jam, and milk. Any news, however exaggerated, was always welcomed. I can see the Australian officers all day long whipping up omelettes for them, while others did all they could to make them comfortable. What could be done was, but that was very, very little.

No more gallant action was fought in the war than that of our sick, desperately sick men, against death. Day after day they willed themselves to live, knowing their chances were almost non-existent. They never complained although they must have felt the lack of drugs and nursing facilities were a direct crime of the Japanese aimed at their individual self.

The food, ration and drug situation was here similar to our previous camp. In order to save more men for the railway, a joint cookhouse was run for both Australian and British camps. Later it was found, when a microscope was made available and clinical examination was possible, that three of the cooks were carriers. Anti-cholera serum promised over a month ago was now just arriving and vaccination was taking place. The Australians were being done as we arrived; we were done next day.

In July we received our first pay since leaving Changi. We were paid at the rate of $68 per month, receiving two months. No deductions except 10 percent for camp funds were made so at last we could buy food. Further, all the other camps down the railway had all made donations to the three cholera camps so now at last something would be done for the sick. Two or three times a week everybody available trekked down to Smoky Joe's, the Thai canteen at the river, and back one would trudge loaded with eggs, tinned milk, peanuts, pig oil, sugar, *gula malacca*, tinned fish, biscuits, tobacco of the native variety known as Wog Weed, and occasionally even tins of strawberry jam. To get to it one had to walk through forest paths to the top of the gorge when one dropped down to the river by a steep winding path through superb jungle. I never tired of this walk.

Smoky Joe himself was a scoundrel of a Thai but he had a soft spot for us prisoners and did all he could to help us. As, in order to save my shirt I always went down in shorts only, he saw my wounded shoulder which gave him great concern and always ensured me preferential treatment.

On one occasion, just as it was getting dark, news came through that a barge had arrived with eggs. Four of us volunteered to go down for them. Much to our pleasure, we found a small British party who were being evacuated next day and who had rustled a cow. We dined with them on thick veg-and-meat stew, steaks, onions, fried pumpkin and sweet potatoes, with bananas and coffee and even a cheroot to finish up with. It was by now nearly dark so we collected our eggs and set off home. We had scarcely accomplished the steep 800 feet climb when a real snorter of a storm broke. I was with Max Winchester, the Australian dentist, and, before long, we were completely bushed. It took us nearly an hour to find the track. Once found, my bump of locality asserted itself and I led the way home.

So slippery was the path that I must have fallen at least forty or fifty times by the time we reached home at 2 am, covered in mud from head to foot and soaked to the skin. Next morning I found that, of the 150 eggs I was carrying, 34 were still whole, but my haversacks were watertight and all the egg juices were there so the whole contents were tipped into a pan and scrambled. Monty, as usual, turned up trumps and had hot water and scalding coffee waiting for me so that I was able to warm myself inside and out and came to no harm. The other two had tried to find their way back by the railway, missed the turning and didn't turn up till 5 am also quite exhausted. But we did enjoy that meal.'

Ironically, at this time, the egg ration in England was just two eggs a month. So Pilkington and his men were ahead on that count. The rice harvest in India that autumn was also one of the best on record, ensuring plenty to go round there. Meanwhile, south in Malaya, everyone had gone into 'business' for themselves, bartering and brokering: the black market had all but overtaken the formal economy.

'At long last we were able to feed up our sick and all day long those who could make omelettes, fried eggs, etc. for themselves and others. The Australian name 'doovering', taken from hors d'oeuvres and meaning whatnots, became the standard expression of this type of cooking. Better still the sick, now that *gula malacca* and peanuts could be mixed with their rice, began once more to take an interest. The tide was turning and the death rate slowly but surely decreased.

As regards we officers, not only were we able to build up our men but we could supplement our diet too. None of the things that we purchased were expensive. Eggs were nine cents, biscuits $2/50 per large tin, fish $1 for ovals.

One evening I was asked to take charge of the night shift as no other officer was available — they were already too ill or too exhausted by the journey. Seventy of the men who had struggled into camp that day were ordered by the Japs for the night shift, too. The evening meal of rice and gas cape stew was served at 6:30; at 7 o'clock I, with Lieutenant Collins, marched our some 100-strong party over to the parade ground in the Australian camp. Here they were split up into parties of 40 each, 25 Australians and 15 British. In all 200 sick men had been ordered for the night shift. Next morning before the night shift would be allowed to knock off work they would have to be replaced, on the working site, with two hundred more. Till the quota was full men would be kept back by the engineers.

What did it matter to them if a man had to work forty-eight hours because the person replacing him had died of cholera?

My report on this night figures in the atrocity report of H Force. At 7:15 a Jap engineer came along with some eight privates and each private took charge of a party. I had command of a mixed batch of men, some 25 Australians and 17 British. The numbers checked, we set off for the working site, a limestone cliff about 1 1/2 miles away. On the way we picked up the midnight meal allowed for these parties. About 7:50 pm we arrived on the edge of the gorge and could just see at the bottom of an almost vertical slope — some four hundred feet down — men working in a deep stone cutting.'

Japanese engineers decided the line should go *through* the

mountain: a tunnel. But due to a shortage of shipping space – the Japanese were losing one ship per day by now – and priority being given over to other war materiel, no construction equipment could be brought in. Instead they called for an open cutting. This cutting was, of course, Hellfire Pass. RAF POW Flight Officer Charles B Knowles' diary dated 22 May 1943 talks of the cutting 'known as Hellfire Pass'. It is the earliest known use of that term.

'Just as we arrived the blasting took place. At eight o'clock we were checked by our Jap; one man had collapsed from obvious cholera and a replacement was ordered from the camp, the sick man being taken back by the party off duty. We then wended our way down a steep path, in some places a series of steps twelve inches high. Arrived at the bottom, the containers with the evening meal, the men's mess tins and coats were dumped, and again the party was checked for numbers. They were ordered to take over from the men working on the cliff face. This they did, those relieved falling in for a final check by their guards before being allowed to proceed home.

For the first hour everything went smoothly. The Jap guard who had been told about, and had seen, my wounded shoulder left me to my own devices. About nine o'clock with darkness setting in, the scene changed. Paraffin flares were lighted and placed along the working face. In their flickering light the men worked, their shadows casting huge and uncanny figures over the rocks. The air was filled with a fine mist of powdered rock which thickly covered the men, turning them into white-looking ghosts. Above everything was the infernal noise of the drills, the coughing of the compressors, and the continuous rumble of falling rock. Down on the track were three light railings with tip trucks which – as soon as they were filled – were rushed to the tip at the double by men in bare feet, running over flinty broken stone.

Into this ghastly picture of an inferno suddenly rushed a huge Jap private called Musso from his likeness to Mussolini. He was notorious for his brutality up and down the line. Without any provocation he leaped on the men with a big bamboo, beating each man across the shoulders in turn. If he found a man bending down, he brought his bamboo down across their shoulders with the full strength of his two arms. Many men were felled to the ground in this first onslaught, only to be kicked and again beaten till they had resumed work.

Tired out with this little show of temper, he came to where I was superintending the loading of the tip trucks. The other Jap pulled back my shirt, showed him my wound and said, 'This *shoko* no work'. Musso looked at my arm and gave me three savage blows across the head with his fist, shouting, 'One arm no good-ah. All men work two.' He then passed on.

Half an hour later he came back to where I was trying to help my men get a truck back on its track that had jumped the rails. He immediately leaped at me yelling, "One arm no good-ah", and then proceeded to beat me up as savagely as was possible, hitting me over the ears and punching my face. On his first attempt he hit me with his bamboo it broke, which made him furious and he started to kick my shins. Finally he picked up a shovel and hit me over the shoulder with it. When he left, my ears were trickling blood and the right eardrum was broken.

About two hours later I asked permission for a sick man, obviously suffering from cholera, to return to camp (Japanese *speedo benjo*). The only answer I received was a beating up as savage as the first. From then onwards Musso sought me out whenever anything was required, accompanying his requests with savage blows.

At midnight I was allowed to take 10 men at a time and give them their evening meal. Twenty minutes was allowed for this. On their return he checked the party and, with blows from his bamboo, sent them back to work. About 4 am he checked the numbers and found six men missing. He made me accompany him in his search. The men were found lying by the cutting too ill almost to move. With blows and kicks he got them to their feet. I remonstrated, and for my pains received a few more clouts. He then proceeded to kick the sick men in the stomach. Satisfied with this, he allowed them to fall out. All the men I believe have since died.

At 6 am my request to allow two men to return to camp for the breakfast meal was received in silence but, as I turned round to go away, he picked up a crowbar and hit me over the head. Only my thick felt hat preserved me from serious injury.

All through the night his sadism never slackened. Men were

felled to the ground for nothing, hit with crowbars, spades and bamboos. Kicked and heavy rocks thrown at them. Lieutenant Collins, while trying to assist a man to his feet, was hit with a crowbar over his head, raising a lump the size of a cricket ball.

I have never seen or read of such calculated brutality and sadism as was shown by Musso that night. On no other working gang that I was on was anything like this shown, and words beggar description of the appalling horror of that night. All the remonstrances of myself or my officers only brought further trouble to ourselves and our men.

At 7:15, in the grey dawn, the charges were due to be blown and I was ordered to get the men into a wooden culvert, V-shaped, some three feet high and four feet wide; 42 men had to be squeezed into a space scarcely capable of holding 20. When he arrived to go in himself he said they were not far enough in and started throwing rocks in. Still dissatisfied, he crawled in behind me with his paraffin flare with which he tried to burn my legs. I let out a savage kick, which knocked the flare from his hand and put it out. Immediately the charges had been blown the men were rushed back to work.

At 8 am the day shift took over. One man of my shift was found to be missing, only his mess tin and coat were still there. I was ordered to send three extra men down to make up for his absence.

I was by now nearly deaf and so ill with the treatment I received that it was all I could do to get back to camp. Here with Monty's help I got to bed and for three days lay quite exhausted and ill. Major Kevin Fagan, the Australian surgeon, examined my ears, found the left one badly bruised and the right one ruptured but had nothing to treat it with.'

Fagan, of the Australian 2/10, was an outstanding scholar, twice dux of his alma mater, St Ignatius College, Sydney, where he won a scholarship to study medicine at Sydney University. He became superintendent of Hobart General Hospital in 1937, before joining the Army Medical Corps and being posted to Singapore with 8 Division. He was credited with doing over 40 operations in a day (including amputations and skin grafts) on the Railway.

Perhaps Musso's rage was a reaction to MacArthur's launch, on 1 July 1943, of the Allied offensive in Pacific. In all, 68 POWs were murdered by bashing in the Kanyu cuttings area.

At this point, still nothing was known by the outside world of the POW camps in the jungle or the pitiful condition of the men. Malaya and Thailand were completely secreted, cut-off, isolated. Would rescue, if it ever eventuated, come too late?

Meanwhile, on 6 July Tojo Hideki appeared on the dais in front of the Municipal Building, Singapore, watched by a 'rent-a-crowd' of 70, 000 assembled on the Padang. Although many actually feared they were going there to be massacred, all were given Rising Sun flags to wave. In Changi Gaol, 3458 civilian internees still languished.

'At the end of five days, feeling somewhat recovered, I took charge of the Jap ration party for four days under a pleasant Jap. This consisted of shepherding a party of 40 men to the river four miles away, picking up stores and bringing them back. Realising that I was to be trusted, he never molested me or my party, but augmented our midday meals with some Jap meat stew. Very acceptable. He also allowed me to buy 1000 eggs for the camp to be carried up by the party, I giving him 10 percent of the eggs for himself. That was the recognised fee.

Three days later the shortage of officers was still more acute and I was asked to go down with the night gang again. In the meantime my case had been reported which had even shocked the Jap officer in charge who had himself beaten up Musso. Things, however, were little improved and although I came off a little lighter, Major Fagan reported that it would be criminal for me to go near such Japs again. Monty and the Australian Colonel, with the British interpreter, took up my case with the Jap sergeant in charge of the camp with the result that I became a Red Star man, one of those who were only allowed to leave the camp under our own camp guards. My treatment from then on was always kind and considerate.

You should understand that although my case was one of sadistic brutality, it was by no means uncommon. Other officers had their arms broken, were knocked unconscious when tying to remonstrate, and I know of no officer of a working party who, at some time or other, wasn't savagely beaten up.'

Interestingly, but perhaps not surprisingly given other

bigger cruelties, Pilk makes no mention of an episode he suffered at the hands of an increasingly drunken 'Nip Quartermaster' which appears in *One for Every Sleeper* by Major Jeffery English: 'I myself was huddled under a blanket having a malaria rigor, and he took no notice of me; but the man next to me, by the name of Pilkington, had a sorry time, being beaten about the head and then pulled off the *chung* by his moustache into the aisle. This, to a British Captain from a little rat of a drunken nip private: and there was nothing whatsoever that we could do to prevent or to avenge this or a hundred other similar humiliations.'

'One of the worst features of the Jap administration was that at these camps no special carrying parties were allowed to draw our rations from the river dumps. This meant that men who had worked all night were only allowed four hours' sleep from 9–1 o'clock. They then had their tiffin and had to go down for rations. The average weight that each man had to carry back would be some 60 lbs. They seldom got back before 5 o'clock, had their meal at six and at 7 o'clock were on their way to another night shift.

It was a programme of deliberate murder.

From the time I became a Red Star man I devoted my working hours to the hospital and in particular to the cholera and dysentery wards. Every morning I made my rounds giving the news and generally trying to cheer them up. Later I took them their share of canteen goods, cigarettes, etc., distributed eggs, sugar, *gula malacca*, etc. During the daytime I would try to fulfil their little requirements in the way of shirts, blankets, mugs, etc., which I would obtain from dead men's kits, and distribute after sterilisation. It was a harrowing experience. Every day some man who you thought had turned the corner would die. Always the death certificate would say 'Starvation and heart failure following cholera (or dysentery)'. Never once did our doctor state anything but the truth.

Day by day you would see men with dysentery gradually fade away till they were nothing but transparent skeletons. They all died with a smile on their lips and a word of thanks for what little one had been able to do for them. The doctors and orderlies were superb. Never did they falter.

Tropical ulcers were now a serious menace to life. Nothing

would cure them. Bandages had long since run out and only leaves were available. Major Fagan eventually received complete operating equipment and anaesthetics. He was then able to clean the wounds cutting away the diseased flesh. Recovery after that was slow but sure.

At first all cholera cases were cremated but later buried in communal graves in a little cemetery about two hundred yards up the road. Here was erected a big wooden cross, hacked out of a tree with a hatchet by Colonel Oakes, on which was carved with a scalpel "To our British and Australian Dead". Before we left, a padre came and consecrated the cemetery. In all some 50 percent of the men that came to that camp we buried there.'

In a mere 10-week period, 216 deaths were recorded at Malay Hamlet.

'Day and night shifts of 12 hours each were enforced till the middle of July when some 2000 other POWs drawn from Saigon, Bangkok and the lower reaches of the railway arrived. Shifts of eight hours were then introduced.

Only on such occasions, as when the Japs brought in some 500 Australian POWs — who had been living on the fat of the land in Saigon — to help us, was our real state of emaciation and weakness really apparent. These men, meat-fed, were like a miracle to us. Two of them would move a boulder that eight of us hadn't even been able to rock.'

Rails had been laid up to Kin Sai Yok at the 172 kilometre mark by 30 July 1943, which doesn't tally with Pilk's following entries, as Malay Hamlet was at the 153 kilometre mark. However, this is probably explained by the fact that the Railway was built simultaneously from two ends, with other parties floating progressively in between.

The cholera epidemic officially finished on 22 July. After vaccine inoculations were brought in, only three more died of cholera. Still, in August 1943 alone, 1600 POWs perished working on the Railway (with over 1400 in each of the two preceding months at the height of the 'Speedo' period). The 2 i/c of Malay Hamlet, Major A.E. Saggers of 2/4 Machine Gun Battalion said of the time: 'Never have I dreamt that I would see the day when human life would be held so cheaply.'

'At the end of July the Japs stated that they were going to evacuate the worst of the sick. I was to go with the first party. However, no one left till the middle of September when I took the *last* party of British sick down.

Twice during our time here the 'Gestapo' [*Kempeitai;* secret police] turned up, but each time we were warned well in advance by our guards, so all things like wireless and incriminating documents were well hidden away and they retired disappointed.

In early August, with the railway rapidly approaching, we got supplies of tinned fish and milk. In the middle of August, when the rails had actually been laid through our sector, we began to get a minute quantity of vegetables. Beyond stealing our sugar ration, a great deal of our meat and some of our tinned supplies, the Jap QM allowed us all the rice, dehydrated potatoes and gas cape we could eat. Fish became plentiful. The worst feature was the downright and open theft of our meagre salt ration, which was essential for saline injections to cholera patients, to cure cowhides which they hoped to sell for their personal profit. They never did, and eventually they were burned and no doubt the Japs responsible felt no qualms at the number of men they had murdered through their robbery.

Cooking with the open *kawalis* was always monotonous, pap (porridge) of rice for breakfast, stew and rice for the other meals, with boiled water to drink.

Clothes were always a problem. Shirts were seldom worn so that the sweat and continuous washing should not rot them. The result was that we nearly all had an almost respectable shirt, even if the tails had been cut off to patch the rest. Shorts, however, had to be worn although latterly many men preferred to work in loin cloth, like a native, rather than damage his nether wear which he would put on for march time.

It is difficult to look dignified or stand on your rank, as even our colonels had to do, with naked sterns when their shorts had split so much that the legs were attached by only a few inches of cloth in front to the uppers.

The elephants living below our camp caused us endless joy, and envy at the wonderful dried cakes – each the size of a stool – the elephants had for their supper. This resulted in the mahouts often giving us presents of fruit and delicacies and, if they were in a particularly good mood, they would get the elephants to touch us with their trunks.

Every few days Thais herding cattle would pass us to the Jap camps up the line. I have never known the Australians fail to abduct at least one beast, when at night the camp would be lit by millions of little fires over which men squatted frying steaks.

I once witnessed a fight between a small hamadryad and a lizard. The snake eventually won, only to be killed by me.

The jungle life in these highlands of Thailand was entrancing. Bright coloured jays, birds with every imaginable type of plumage, gaily coloured robins and starlings, while on the ground strutted the most beautiful pheasants and jungle cock. In the background, playing round the camps, were many types of monkeys and it would amuse everyone how I could foretell changes in the weather by their different cries. Never have I seen such huge and lovely butterflies, some of them like black lace, others every colour of the rainbow.

To watch these birds and animals as we did with our tired and disillusioned eyes, was to be transported on a magic carpet to a world of creatures that had solved their problems, and were now living in a jungle Paradise.

On many occasions I was able to talk to the Tamil working gangs. They were pathetically glad of a few words of encouragement. They had been collected from estates, with the threat that if they didn't go, no food would be allowed their families. They were rotten with fever, dysentery and sores. They had no medical officers. A coolie sick with malaria was allowed one grain of quinine for five days. Irrespective of the size of the sore, they were allowed a piece of lint about 1 ½ x 1 ½ inches per week. If they were too sick to work they were herded into huts where they got no attention whatsoever. If they died, that was that. If they lived, back to work.

On one occasion they came to me when I was on my way to the river and said, "*Dorree*, would you like a pumpkin? We have just stolen some from the Japs." I was led to a huge pumpkin weighing all of thirty pounds hidden away near the track.

Next day I sent Bowman down with two men to collect it. Back in camp we divided it up, ¼ to the carriers, ¼ to Bowman and the rest to myself. We made a huge pumpkin stew on which we all gorged. There was still sufficient over to feed Monty and myself with fried pumpkin for three days.

Over the period in this camp I fared about the same as the rest. I had three severe attacks of bacillary dysentery, the eternal tropical enteritis never left us and, altogether, I must have had malaria four or five times. My right ear, ruptured by Mussolini, gradually festered till at last Major Fagan and Doctor Moss were able to get some instruments to deal with it. I was by then in a high fever with a temperature (taken by Monty of 103.5). Major Fagan dosed me with masses of M&B tablets, while Doctor Moss dressed it every few hours with Epsom salts, also so rare that the small hoard was a deadly secret. Others kept me warm with their blankets and water bottles filled with hot water. All this time Monty nursed me most skillfully, tempting me with little dishes he had cooked. Gradually the inflammation subsided and, at the end of a fortnight, I was all right.

Fortunately for me about the beginning of August some of our heavy baggage left at Tarsao arrived by barge. My bedding bundle survived the thieving Dutch, Australian and British hands through which it had passed and, much to my relief, I got my blankets, greatcoat, spare pair of boots, mosquito net, pillow, two or three good books, a tin of Lipton's tea, and a large one of my salt I had manufactured at Changi. After that I lived in luxury.

Monty's valise also turned up intact complete with tool bag full of nails and wire as well as hammer, pliers, etc. All the other officers' kit had been lost or stolen, almost certainly the latter. You cannot imagine what a luxury it was once again to walk about in boots with soles on them and that didn't leak.

Boots and shoes featured very largely in our lives. A good pair of boots were worth their weight in gold but up-country they were life. Men waded naked to the waist and barefoot through the gluiest mud in the world, with their boots and shorts slung around their necks. Each step when one's leg sank up to the thigh in mud was a Herculean effort. I spent a full hour prodding the deep mud with poles until eventually we located a man's boot sucked off by the mud, when the proud owner stripped himself naked and plunged up to his shoulders to retrieve his footwear, to return to camp naked but for his mud-plaster covering.

There was practically no chance of getting boots repaired in Thailand, so once the sole was off 'you had had it'. Hence the marvellous repairs with wire nails and string and the look of despair when finally the sole and the uppers parted company.

Dead men's boots would be stolen almost before the man was dead but latterly, owing to our very high death rate, the boot situation improved.

The cuttings were completed about the middle of August when, except for a small day shift, work on the railway for our camp terminated.

We few survivors of cholera, dysentery, malaria, starvation and brutality were too weak and useless for the Nips to worry about any longer and, at long last, we were allowed to rest and recuperate as much as our diet of rice would permit.

As regards the food in this camp, up to the middle of July it was nothing but a repetition of our previous camp, but then, thanks to the representation made by Colonel Humphreys, our 'H' Force Commander, food began to arrive. At first half a dozen pigs, then twelve cattle on the hoof, which we corralled and slaughtered as permitted. Later, we had for the last three weeks so much meat that we could hardly eat it.

I used to assist pathetic little convoys of the very sick down to the river at Kanyu, mostly of men in the last stages of beri-beri, with their feet and legs so swollen that no shoes and socks could possibly be put on, and their only hope of life was to trudge the six miles to the waiting barges by the river. One, suffering from beri-beri, was gallantly cheering on his own companions although, literally, all one could see from his bloated and swollen body were finger and toe nails. The rest of him was one vast balloon. The cemetery by the river — with its rows of crosses — was the last answer for most of them. The river once made, the waiting barges promised by the Japs would not materialise and, finally, despair and disease would conquer their gallant souls.

If medical drugs had been supplied, the food kept at the standard when we left, and a proper water supply installed, we would all have really loved this camp, but as it was its beauty was marred with the cruelty, starvation and deaths of our companies.'

The line had reached Tha Khanun at the 218 kilometre mark, work made more treacherous as Kanchanaburi province received about 40 percent of its 7000 mm of annual rainfall in the two monsoonal months of September and October 1943.

'Those that died on the Railway, and they were over 50 percent of our party, never gave in. I can remember on one occasion talking to a man who in turn was helping his neighbour. When I came back an hour later and asked for him, he was awaiting burial. One and all, with their last strength, assisted the nursing orderlies who were looking after them. None of them seemed to suffer any pain and, in spite of the squalor and misery of their conditions, they all died peacefully. Gradually they drowsed off into the coma from which they never woke, with a smile, usually a very happy smile, on their lips.

The verbal last wills and testaments given to me by those who knew they were dying were a great tragedy, instructing me as to how their personal possessions were to be divided amongst their friends. The heartrending business of going through a dead man's kit and destroying all his papers and photographs that he had so cherished and which, if the means had been available, one would have returned to his next of kin. The dead man's mess kit, razor, clothes and blankets, which when boiled, could be handed over to the most needy. Sometimes there was a photograph of the man himself in England, when, comparing the stalwart well-fed picture with the shrunken starved corpse at one's feet, drove home to me the full enormity of the Japs' criminal neglect and cruelty to us.

Death to some extent is infectious. If one man of, say, three friends died, the others very often did too although there was no reason for them to do so, and very often they would die within a few hours of their friend.

Owing to the fact that every one of us was living on the same rations and usually suffering from the same disease together, our gradual loss of strength and weight was not really apparent to us unless we kept a weight chart, which few of us did.

We had nothing except our memories to contrast our way of living with. Hence I think we were able to bear our lot with much greater fortitude and the smallest improvements in our rations or our comforts were to us very greatly magnified and helped us along considerably.

There can be no doubt that if anyone kept an animal in England in the conditions in which we lived and our men died, the RSPCA would immediately prosecute with a certainty of a maximum conviction in the courts.

Grim scenes at Malay Hamlet, adjacent Hellfire Pass. Above: the kitchen. (AWM 132943) Centre: The dysentry block. (AWM 132941) Below: The cholera block up the hill. (132944)

The Wang Po Viaducts, an epic engineering feat cut into the limestone cliffs. (AWM 122323)

Right: The two ends of the Railway were joined at Konkoita on 16 October 1943. (AWM P00406_004)

Kanburi, one of the largest hospital camps. (AWM P01502_006) Tarsao was considered one of the foulest camps on the Railway. 1000 POWs would die here. (AWM P01910_005)

A typical work party on the Thai-Burma Railway laying tracks. (AWM P00406_027)

The bridge over the 'River Kwai' bombed by the USAF in June 1945. (AWM P00502_001)

14 ONE STOP NEARER HOME

Malay Hamlet to Kanchanaburi Hospital Camp, Thailand, 14–21 September 1943

'On 9 September orders finally came through to evacuate all sick from the camp. On the morning of 10 September the first party of eighty left. With this went Monty. How on earth that man succeeded in carrying his valise, all his other gear, plus his beloved tool bag, I don't know. It was all I could do with the help of another to get it on to his shoulders. This party contained no stretcher cases.

Another party went off on 11 September and then again on 13. We were now getting the worst cases away on stretchers constructed out of bamboos. To take the place of the canvas we fitted bamboo slats. In our weakened condition it took four men to a stretcher.

Everyone was being evacuated by construction train from the cutting about 1 1/4 miles away [possibly Hintok, two kilometres north of Malay Hamlet, being the nearest station]. The road to it lay past the Jap HQ when, in order to avoid the main track which was thigh deep in mud, a path was cut through the jungle avoiding the worst part. We then turned off through the camp that had recently been occupied by some 60 elephants, whose antics, cleverness and good nature had given me many hours of amusement, past the Tamil camp and then down a steep path to the railway. A normal person could walk this easily in 25 minutes but our walking wounded with their swollen beri-beri legs often took an hour and a half. One man's legs were so swollen that wearing boots was quite out of the question. From his thigh to his feet his legs were such a size that all one could see was a large very fat limb with about three inches of swollen foot sticking out. All shape and resemblance to a human being's leg had been lost.

Monty gone, I took over the administration of the British group. As each party went off I, with every man that could assist, did what we could to help. The strongest carried stretchers, others the kit of the worst walking cases, and still others helped these cases down.

On the morning of 13 September the Japs said that all that remained of the British and Australian sick would be evacuated next day at 6:00. The total number was about 96 of which 26 were British, nine stretcher cases and 15 walking. The rear party of all our fit, some 60 including both British and Australian, were to be left behind to strike the camp. I left Bowman with 15 men, all that could be called fit out of 400.

At 5:30 next morning we fell in, the Australian officers and padre as was customary falling in as stretcher bearers, the Colonel going with the first stretcher. Two of my men carried my belongings down for me. At 6:00 the party moved off, looking more like a cortege than anything else. By nine o'clock the last man arrived at the railway. Here we sat till 11:30 when a construction petrol engine came along shoving two trucks.

These engines with diesel motors were most ingenious as they were capable of running both on road and rail and were very powerful. Into these trucks we crowded, there was now no question of people lying on stretchers; the most desperate cases had to stand or sit, if room could be found. Normally with care each of these trucks could hold, if carefully packed, thirty as a maximum. They now had to take fifty sick men each. Everybody aboard, we moved at about five miles an hour over the railway we had helped to construct, finally arriving at Kanyu siding some six miles down the track at about 12 o'clock. Here we were turned out but, for once, found that tents had been erected to give parties shelter. Lunch of rice and dried fish was served.

At two o'clock an empty construction train, which had been unloading, came up to our siding and we scrambled aboard — this time there were only 12 to a truck but no shelter from the sun. For four hours we moved slowly down the track, over bridges that looked and felt as if they would collapse at the least encouragement. The scenery, as always round here, was magnificent in its stark rawness of limestone cliffs, jungle, precipices and river winding hundreds of feet below.

Occasionally we passed huge waterfalls and always gangs of POWs working on the line with whom we exchanged greetings. It was pathetic to see their looks of envy when they heard we were going down, but one and all wished us the best of luck. About four o'clock it poured and everyone got soaked. At 5:30 the sun came out and we dried off quickly in its heat.

We reached Tarsao at 5:30. Here we were ordered out and told to make our way to the new transit camp some 1 1⁄4 miles down the road. By great good luck I saw one of my own men driving a truck into the station. Delighted, he rushed across to see me. When I had explained the position he immediately went off to his Jap sergeant who came up, looked at our sick, and without any hesitation ordered all his twelve trucks to be put at our disposal, an act of kindness so rare amongst that race that I record it with gratitude.

The transit camp proved to be half a dozen V-shaped *atap* huts with a row of sloped earth banks on either side to act as beds. Here I found that the majority of the stretcher cases already sent off had been detained, some by the Jap failure to supply sufficient transport, others at the orders of the British MO in charge of the camp who, on examination, didn't dare send them further in their weakened condition. Everybody housed, a good meal of rice and thick vegetable soup and boiling hot water was served.

Under my especial care I had Captain Shakespeare of the Signals, desperately ill from malaria, who a few days ago had not expected to live. He did, however, survive the journey and was now fit and well, but, added to my responsibilities with the sick, it meant that I got no rest which by now I badly needed.

That night we were informed by the Japs we should be there some days so I went in search of my old friends Colonel Knights and Padre Alcock. They were, I found, as delighted to see me as ever and soon I was bathed and shaved and drinking coffee. For the next four days they gave me meals and did everything they could to help, but my request that the Padre be allowed to visit the transit camp was turned down by the Japs with a firm "No". Those in transit were, so to speak, in *purdah* [seclusion]. No one must touch or see them.

Canteen parties were soon organised and I toured the little town buying eggs, fruit, toffee, *gula* cakes, biscuits, limes, etc. Nearly all the canteens there were being run by the various construction parties for their own men but, the position explained, they always gave us all they could spare and generally a lot over. One Dutch officer in charge of his canteen gave me 50 eggs as a present for the sick.

With these canteen supplies I hoped to build up the sick and undoubtedly they did help but, for many of them, the journey had been too much and they died. In all nine men died out of all the British stretcher cases I had with me here. It grieved me considerably as we had hoped now that we had nursed them over their illnesses and they were on the way to better things, that they would live. Their beri-beri hearts, however, just couldn't do it. They were all buried from the hospital by Padre Alcock with Colonel Knights and me as mourners.'

Tarsao Hospital Camp was adjacent the Kwai Noi River, slightly west of Tarsao Work Camp.

'It was nice to see the dead at last being buried with a little pomp and ceremony. As the cortege passed everyone got to their feet and stood to attention, very different to the indifferent looks the dead received in the construction camp. At the conclusion of the service the usual bugle calls were sounded.

While in Tarsao Colonel Knights got permission for me to go round the hospital wards of the British Hospital which at that time contained some 2000 cases. Padre Alcock provided me with a list of all my men that were here as well as a guide. I had more of a triumphal progress than that of a decorous officer condoling with his sick. The men turned out to greet me while amongst them were many from other units that I had had dealings with. The magnificent work of Colonel Knights at Tarsao and Padre Alcock is one of the shining lights of the railway camp. Even the Japs had such respect for Colonel Knights that their soldiers were ordered to salute him, receiving a severe beating up if they failed to do so.

Midday of 14 September orders were given for a party to move off at two o'clock for the base hospital. No stretcher cases were to be taken. I was by now suffering from dysentery and malaria, which up till now I had been able to suppress, and was ordered down immediately by the MO. My request to stay with my stretcher cases was firmly over-ruled. One of my men in Tarsao met me at the entrance and he and others carried my kit to the station. Here we were loaded into covered cattle trucks which by now appeared to be almost first-class travel.

I shared a truck with Shakespeare, some ten soldiers and three Australian officer friends. At six o'clock, after three hours' wait, the train moved off, darkness soon descended and we rattled through the night.

About midnight the train stopped at a hutted camp, we were told to get out and the train moved off. Almost immediately a horde of orderlies and others descended on us, seized our belongings and helped us to the entrance where we were sprayed with a carbolic mixture (as a precaution against what I don't know, as the stuff never ever made a louse blink). After being paraded and checked by the Japs inside we were led to newly erected huts. Here we were given our bed spaces on bamboo covered platforms. For about the only time in my period of imprisonment I insisted on my seniority and took the corner space.

A few minutes later Monty and Colonel Humphreys dashed in, Monty delighted to see me. Almost at once a huge mess tin full of rice and thick meat and vegetable stew was served to us boiling hot, with big mugs of hot tea. Monty also had some hard boiled eggs which he cracked and peeled for me so I ate and we talked. I was famished and I shall never forget those eggs.

The meal finished, lights out was ordered by the Japs but at that moment Jacko came in with six letters for me from Phyllis that he had sorted out of some mail that had arrived. I just had time to read the postal dates and make sure it *was* her handwriting when darkness and silence descended on us. That night I slept better than I had done for many months.'

Meanwhile, 17 September saw Roosevelt and Churchill reached agreement on the times and places for new offensives blows to be struck in Europe and Asia. A couple of months earlier the USA had also agreed to share its latest promising research with Britain: something called an atomic bomb.

'It really did seem as if home was getting nearer.'

15 THE GRIM CIRCLING OF VULTURES

Kanchanaburi Hospital Camp, Thailand, 21 September–14 December 1943

The Japs, at last appalled and frightened by the ever mounting death roll, had ordered this hutted camp to be constructed for the survivors of F and H Forces. It was situated within a few yards of the railway in flat country that was mainly paddy fields. The huts were of Thai pattern, some 120 feet long and 30 broad, with *atap* roofs and mat sides. Down each side ran the bed platforms of bamboo and bamboo slats over which were laid thick grass mats. Down the centre was a passageway some six feet wide. The huts were good and serviceable. At first we all got a bed and living space of about three feet wide by 12 feet long but, as more and more sick arrived, this was gradually reduced to 20 inches each.

In the centre of these rows of huts, some 24 in number, were the kitchens while flanking them were deep trench latrines. Surrounding the camp was a tall bamboo fence. Outside the huts when it rained it immediately became a gluey quagmire of the sticky variety that only Thailand could produce. Fortunately, however, after a few hours sunshine this dried up into hard firm mud.

If the facilities for cooking had been better we would have lived very well.

The main difficulty was water and wood fuel. The former had to be brought by trucks from the filtering plant by the river, four miles away, while the latter had to be brought in from a considerable distance. For washing purposes water was nearly unobtainable so that the only baths we got were from the rain sluicing off the eaves. It rained most nights and, as we were able to purchase buckets, we generally had sufficient to keep moderately clean. Some Air Force lads that I had befriended turned up trumps and kept me supplied with water that they got from a well some distance away.

Under the confined conditions in which we were kept, without adequate water, it is not surprising that lice and bedbugs, from which up to now we had been free, should make their appearance in swarms, with the result that before we left that camp everyone had scabies.

No one here could forget the three o'clock scabies round of the medical orderlies, when every officer and man in turn down the long hut beds heaved himself to his knees and presented his stern to the orderly so that the scabies places could be anointed with iodine.

One afternoon the orderly was suddenly called away as he was about to deal with an Australian major so I, unknown to the major, took his place and painted with iodine a beautiful head, nose and eyes on his seat!

Canteen supplies were good and cheap. Everything we required was now available and we were able to augment our rations. I used to mix four spoonfuls of sugar with my breakfast pap and then squeeze three or four limes over it. At 11 am my cook would appear with a four-duck-egg omelette and again at four. For supper at six o'clock I would have a piece of pomelo, some biscuits and peanut toffee. If possible, acting under medical advice, I would eat, in one form or another, eight eggs a day. In the circumstances it is not surprising that most of us soon began to shake off the ill effects of the past.

From now onwards the Jap administration did try to do something for us. Quantities of fresh vegetables poured in, pigs were allowed at the rate of about one to 250 people per day, tea was supplied, salt was again available, and all the rice we could eat. A day's menu would now be a pint of good thick rice pap (porridge) for breakfast and ½ pint of tea. Lunch: vegetable stew and rice. Dinner: meat and vegetable stew, one pint of tea. Every third day we got a fried rissole.

One of our men had succeeded in knocking off one of the Jap guard's chickens. This he had plucked and trussed but then found no means of getting it cooked. He therefore offered it for sale but, as at this time there was a strict camp rule that there was to be no cooking done for private orders in the cook house, none of us were able to buy it. When, however, the chicken was offered to a Very Senior Officer, he promptly gave orders for it to be cooked for himself for his evening dinner.

The fury of the rest of us was great, and it was decided that even if we could not stop him having it cooked, we at any rate could prevent him enjoying it.

Dinner was served and a well-roasted chicken, complete with sweet potatoes, was placed before the Very Senior Officer.

1st officer: "Has anyone heard if that Australian was able to get rid of the vulture he was trying to pass off as a chicken?"

2nd officer: "I cannot make out what happened to it, it has completely disappeared. Anyhow I think the Senior MO should put his foot down. After all, one must draw the line ... and *vulture*!"

Another officer: "I heard he knocked it down with a stone coming back from the funeral last night".

Yet another officer: "Oh, no, it was sitting on top of the latrine shed".

And another officer: "I don't suppose vulture is poisonous or dangerous, do you? But I should imagine it would be pretty rank."

General chorus of officers: "Well, I think anyone who could descend to eating a vulture must be as low as a snake's belly!"

The conversation continued until the Very Senior Officer departed, his chicken hardly tasted, when those present fell on it with avidity.

For some, however, only too many, these improved conditions came too late and every day at six o'clock a sad little procession of Union Jack-covered bodies with the Padre at the head, would wind out of the camp. Some days there would be nine, on others only one or two, but only too often there would be ten or twelve.

There are two thousand survivors of the healthy six thousand that passed this way six months ago.'

Tragically, 550 more sick members of F and H Forces perished from their irreparable privations after they'd come down from the Railway and gone into the hospital camp.

As the expression goes: 'There are only two kinds of people in the army: medics and patients'. Never was this truer, nor more patent, than on the Railway. Around 350 Allied doctors had been taken captive. One RAMC doctor, Canadian Major J. Markowitz, was the sole surgeon for eight months at Chung Kai Jungle

Hospital. With neither instruments nor drugs at his disposal, he was in charge of 7000 desperately ill POWs, becoming known as the 'Knife, Fork and Spoon Surgeon'.

'But the Japs were by now pouring quinine and Epsom salts, bandages and lint, ointments and other medical necessities, most of which had been looted from our Red Cross supplies, into the camp. They also opened a black market for emetine, the life-saving drug for Amoebic Dysentery, a course of which cost $120. M&B 693 was also procurable. An operating hut was built, within which miracles of surgery were performed by Major Fagan, who was operating daily for four to five hours at a stretch on the tropical ulcers. Many people had ulcers that had gone too far for treatment so that amputations became common. Nevertheless they are now alive and well.

The day's routine was breakfast at eight o'clock. MO's rounds 9:30–10:30 am. Lunch 12:30. Tea 4 pm and dinner 6 pm.

No light other than one lantern for each hut was allowed after dark and this had to be shaded at 12 o'clock so that it couldn't be seen from the outside. This was due to the cheering sight of British bombers flying over to knock a bit of Bangkok down.'

One of the largest industrial concerns in Kanchanaburi was a paper factory. Its lofty chimney proved an excellent beacon for Allied bombers.

If there was a glint of optimism locally, it flared internationally. By the end of October the USSR, Britain, USA and China met to sign an accord on the post-war attitude to Axis. Still, conditions on the ground remained dismal ...

'To take the place of books and pass the time between 6:30 and 10 pm when it was dark and the single hurricane lamp could do no more than act as a guide down the huts, endless quizzes were arranged and competitions on the 20 Questions principle.

In the evening we took it in turn to lecture to the various huts on every subject under the sun, held eternal quizzes and, occasionally, a song or concert with Flight Sergeant Williams at a little organ that had been procured.

Literally it got so cold at night that I slept with all my clothes on including my greatcoat, rolling myself in my blankets like a cocoon, finishing off by wrapping my mosquito net round me. Even then I would wake up shivering with cold. These nights outside were nearly up to the standard of a frosty night at home.'

On 30 September Phyllis sent to Pilk a photo of an eighteen-month-old Paul standing up with his bare behind. On it she plaintively scribbled: 'Please, censors, let my husband have this picture of his baby he has never seen. Darling — this gives an idea — taken July. Started walking then always runs! Loving P.' The photo was duly returned, along with a letter from British Postal Censorship:

'Returned to sender by the censor: This photo is returned to you for the reason underlined 1/ Because it might be useful to the enemy as it contains a/ information about the services b/ economic information c/ matter useful for enemy propaganda d/ reference to air raids. 2/ Because it contains abuse of the enemy, which would probably cause the detention of your letter, and might be detrimental to the addressee of the letter.'

No reason for its return was underlined.

'From my corner bed I can see the telegraph wires of the railway we have constructed with dead men. I think the period we were here must have been the migrating period because I have never seen so many different birds in a short period. They loved to sit on the telegraph wires by the railway. The few durian trees that break the monotony of flat green paddy fields are packed with grim-looking vultures wanting to pay their last respects to our dead when they form a great aerial cortege, following the quiet group of figures that are carried each day to our cemetery half a mile away.

The paddy fields at times were almost white with egrets. In our hut some sparrows built their nest and brought up their family, living on the little titbits we put out for them.

But as the sun sets and the last splash of red is in the sky, the telegraph wires spring to life as with a magician's touch. Here beauty and colour was unmatchable, starlings of every description, swallows, swifts, parrots, starlings, kingfishers and fly catchers. All chirpy and happy, all in love with their world.'

Meanwhile in the horror and disorder of the human world, October 1943 saw Mountbatten became Supreme Commander South-East Asia, with approximately a million men at his disposal in SEAC. Japan had roughly equal numbers deployed throughout South-East Asia.

The Japanese ceded four northern Malay states to Thailand – Perlis, Kedah, Kelantan and Trengganu. Coincidentally or otherwise these were the rice bowl of Malaya. 'Whoever controls the rice-pot of Malaya, controls the destiny of the country,' wrote Chin Kee Onn in his memoir *Malaya Upside Down*. Placatory concessions were made to the Malay sultans, acknowledging them more than in the original plan to diminish them under Marquis Tokugawa's scheme but, still, they were handed this as a fait accompli by the Japanese.

'Microscopes being available and proper clinical observation again possible, Mr Pereira, an Indian doctor who I was friendly with, promptly took me in hand as he was running the laboratory, and he diagnosed BT malaria and bacillary dysentery. With proper, or as near proper, diet and a full course of English quinine supplied to me by Dr Young, I soon began to recover and threw off the dysentery once and for all.

BT malaria was another matter and I had four or five relapses in spite of Major Fagan getting me Atebrin and plasmaquin. Eventually, just before we were due to leave, the staff sergeant who was puzzled by my temperature (we had no thermometers so temperatures were assessed by touch) suddenly realised that I was developing yellow jaundice. I have never felt so depressed and miserable. Starvation and Epsom salts were the cure with a very light diet some days later of towgay bean soup, tea and sago and lime porridge.

The whole of the period I was here I retained my corner, which became known as 'Rumour Corner', as it was just by the main path when everyone passing down would look over the mat wall at me as if I was a horse in a loose box, and give me all the latest news and rumours, which I passed on to all and sundry.

A lot of my time was taken up distributing eggs and little luxuries that we bought with subscriptions from our officers, Daykin and Bidulph, Dr Jackson and others all subscribing. What little we could do in this direction was most appreciated, and I like to think that we saved several lives.

Much to my annoyance by getting dengue I missed going down to the fit camp where the food must have been superb as Monty, with whom I was to go, dribbled at the mouth when later we met and he told me all about it.'

Meanwhile the US Pacific Airfleet bombarded Rabaul on 12 October, wrecking 177 planes and 124 ships in surprise attacks. Four days later, the Aussies repelled a Japanese offensive in New Guinea.

On 16 October, the Burma and Thai ends of the railway joined at Konkoita, at the 262.87 kilometre mark, a couple of hundred kilometres north of where Pilk and his men were recuperating. A ceremony was arranged, with high-ranking Japanese dignitaries from the region, plus freshly scrubbed and outfitted POWs, in attendance. A ceremonial golden spike was driven in to the last sleeper, and the Bangkok to Rangoon railway – that wholly unfeasible engineering feat – was linked (although not fully operational until December). Ceremony over, the POWs' fresh uniforms were taken back from them.

According to Japanese records, 4 million cubic metres of embankments had been constructed. A further 3 million cubic metres of rock had been shifted by hand. And fourteen kilometres of bridge had been erected. All completed at a rate of 890 metres per day.

While this was quantifiable, the human cost was immeasurable.

Officially 12, 800 Allies – British, Australian, Dutch, Americans – lost their lives in its construction, along with up to 100, 000 Asian *romushu* workers and their families – only three of whom received a named grave.

The Japanese originally planned the railway to carry 3000 tons per day in each direction. This was later scaled back to a less ambitious 1000 tons per day, due mainly to the formidable terrain.

'October 30 I celebrated my birthday with Shakespeare whose date was 28 October. We had a huge blowout of Boson's omelette (made with tinned pilchards), bananas, sweet coffee and a meat pie! To this feast I invited my own few friends who had been so good to me, getting my water, doing my washing and producing my little extras cooked at the right time.

With all the eggs and vegetables my eyes were by now, although weak, completely recovered. A blessing beyond description.'

Come November, the Americans had secured a numerical and qualitative naval superiority, advancing steadily across the central Pacific. The tide of the war was turning surely against the Japanese, who now had to be more accommodating to local communities, relative as that term may be.

'The Japs insisted on British guards patrolling with them at night due to the loathing with which they were regarded by the Thais, who took every opportunity to stick a knife into them.

When fit enough, my job was to act as policeman with one other officer from 4 am till 8 am, which I found rather pleasant. I was generally awake by then anyhow and the half-hourly tours of the fences in the clear starlight night were refreshing. In between rounds we made the kitchens our HQ where we got hot tea and fried eggs. If it rained I would report to the Jap guardroom perhaps twice during the time, staying in bed for the rest of the tour of duty. As for trying to catch any of our own people, I let it be known beforehand when I should be about so they could do whatever black marketing they wished.'

Meanwhile, at the end of November, Churchill, Stalin and Roosevelt met in Tehran. France was to be the all-important 'Second Front' in the west, meaning no resources would be available to Mountbatten's SEAC. His burning plan to launch an amphibious attack on Malaya or Burma — Operation Zipper — would have to wait. The stalemate would continue.

Incredibly, the Malayan Planning Unit was being ambitiously and prematurely briefed on drawing up plans for the restoration of British government of that country, detailing everything from power to legal infrastructure. Unfortunately, it paid no mind to the status of the war in Malaya. Nor of the fact that the people of Singapore and Malaya were forever changed. Uncertainty ruled. What would the future loyalties of the Malayans be? While they were often divided along ethnic lines, their apparent abandonment by the British, and struggle at the hands of the Japanese, had forged a collective nationalism.

'On 9 December the Japs announced that all those of H Force who could travel were to be sent back to Singapore.'

16 SOUTH TO SINGAPORE AGAIN

Kanchanaburi, Thailand, to Singapore, 14–19 December 1943

'Originally I had been detailed to go on the first train in order to act as assistant QM to Monty at the new camp to which we were going, but the virulent attack of jaundice which I had contracted precluded this and I was kept back to the last train.

On the morning of 14 December we had breakfast as usual and then at 11 o'clock, while the 'Gestapo' were searching our kits, we had a huge last meal of meat-and-veg stew, rice and fruit. Each officer and man was given four large fried rissoles and $5 from the camp funds for the journey. At 12 o'clock all except a handful of sick marched off to the station some three miles away. I, with my baggage, travelled on a lorry in style. While waiting to leave, staff sergeants and doctors kept coming up to say goodbye, each one thrusting into my hands packets of Epsom salts.

Arrived at the station, an Australian private who I had got to know and who was travelling in my truck, took over my kit for me. We entrained about three o'clock. I was in charge of a steel truck with twenty others beside myself. This was luxurious travel after what we had been through. In fact first class! I found that I had to control some 15 Aussies, two Dutch and three British. However, we all piled in amicably, I taking a seat by the open door so I could look out and get out in comfort.

At four o'clock the train moved off in sweltering heat. At six o'clock we arrived at a railway siding in Nong Pladuk, the new junction constructed for the Rangoon railway, about three miles north of Ban Pong where we had detrained on our journey up. Since then this place had been obliterated by the RAF. Here we found that we would have a four-hour wait.

Much to my amusement, as soon as they heard this, my Aussies were out like a flash, had ripped up some sleepers from the track, cut them up and had a fire going, round which they gathered and started to cook. Their example was followed by all the other carriages. The Japs thought it a great joke!'

By late 1945, only the main train line through Malaya was operational. The Japanese stopped all maintenance, but used the railway heavily to feed the war effort. They gave priority to strategic port lines back to Japan, and dismantled track that would be of more use elsewhere.

'At 8 o'clock we were made to put the fires out and carriage commanders were summoned to a conference. Here the Jap officer, who was quite a good sort, said that —providing no one got out of the train without leave at halts — he would put no restrictions on our buying.

I went back and, knowing my Aussies pretty well, gave them a sharp lecture on discipline, to which they listened politely. As far as I was concerned they could do what they liked providing the few orders I was likely to give were promptly obeyed. We then settled down for the night which passed peacefully, the train eventually moving off about 3 am.

All went smoothly for a bit till one of the men at a halt asked permission to get out, which I refused as it was obviously only a temporary stop. Thereupon he leaped out only just catching the train in time as it moved off. I said nothing till we got to the next halt when I told him to accompany me to Colonel Oakes, the Australian Commander. He tried to bluster so I started to detail two men to put him under arrest. He then accompanied me down the train, looking more and more sheepish. Just before we reached Colonel Oakes' truck I turned to him and asked him if he thought it really was worthwhile disobeying orders. He was most contrite so I told him to go back and warn the others that I was going to stand no nonsense. This he did and the rest of the journey passed in a very delightful atmosphere of goodwill and obedience.

By now everyone was an experienced cattle-truck passenger. It made me laugh to see with what haste civilians and natives, both men and women, beat a hurried retreat from the lavatories as the train came to a halt.

At railway stations orderly queues fell in for a shower under the engine hydrants, orderlies told from each truck robbed the engine of boiling water for tea, everyone in each truck bought all they could and we squared up afterwards.

By knowing the languages, I had been able to buy all the supplies of fruit for the trucks at our end of the train at reasonable prices. I literally bought up whole stalls at a time.

I made out a roster of duties such as three men to fetch breakfast, three to fill water bottles, etc. Food I dished out myself instituting a proper roster for extra helpings. In fact we had a most comfortable journey.

Bickering I stopped by telling them that the only person they could argue with was me and, if they did that, I would put them under arrest.

Halted one moonlight night I was sitting by the track while everyone else slept when, to my joy, a pair of bitterns appeared on the edge of the paddy field not twenty yards away and 'boomed' at the moon. On another occasion one evening a gaggle of secretary birds waddled up and fed quite happily like a drove of turkeys within a few yards of me.

The journey down as regards food was a repetition of that going up except that eggs became rarer and rarer till finally I had to pay 25 cents in Penang for a small hen's egg. These were the last I got.

Malaya looked just the same, but very deserted, shops shut, no traffic on roads and very few people about.'

The Japanese removed 50 percent of the up-country work force during Occupation, and more than 150, 000 Chinese coolies died working for the enemy (according to the *Daily Sketch*).

'At Seremban we had a halt of an hour. A coolie who knew me and who was now a porter recognised me and we had a long talk. In the meantime he sent into the town for Vanamuttu Pillai of the Planters Stores who came and gave me all their news. They were all most unhappy under Jap rule, being bullied and tyrannised with sadistic thoroughness. Food was quite plentiful but of very poor quality, and always likely to be commandeered. I got messages to Sam Yeow and Fong Eck who were both reported well, but the train moved off before they could be got to the station. As a farewell present they loaded me with bananas, papayas, pineapples, bottles of lemonade and sponge cakes.

The first class waiting room where I had had so many

pleasant meals was full of Japs, and sold nothing except cheroots and cheap brandy.

In Seremban bandits stole cars and bicycles, and broke into houses demanding ransoms for kidnapping. Japanese commanders in Seremban burnt piles of cloth and military stores on the local football grounds, possibly the very ground on which Pilkington represented his state. (The locals received one yard of cloth ration in those three and a half years.) Every European residence, probably including Ainsdale, was looted of furniture and domestic fittings such as refrigerators, geysers and electric fans.'

In September 1945 the retreating Japanese looted and burned Malaya before they left. Rubber could not be produced post-war for at least eighteen months, with large tracts overgrown by weeds, and machinery laid waste. Tin mining was also compromised by machinery being ruined or lost. Vast tracts of rubber were uprooted and turned into *kinto* (victory gardens), growing tapioca and sweet potato.

Thousands of rubber trees, worth millions of dollars, were ruthlessly cut down in this area by the departing Japanese in 1945. Still, the Brits themselves came in for criticism for their own 'scorched earth' policy, given they and the approximately 31, 000 European residents of the time, were scuttling a country that was technically not even theirs.

'At 10:30 pm on 18 December we rattled over the Causeway. At midnight almost to the minute we steamed into Singapore station.

How glad we were to see it. The journey down, in spite of discomfort, had been almost in the nature of a Bank Holiday outing.

At Singapore we were met by trucks and whisked off in convoy to the Sime Road Camp. Here a reception party met us, carried our belongings to huts ready for us, and there we were given mats to sleep on. A large hot meal of rice, vegetables and beans was served. I had a wash and shower and — feeling at peace with the world — turned in and slept till 10 o'clock next morning.

Out of the British party we had lost 68 percent dead in squalor and misery. Of the total of 10,300 from F and H Forces only some 5000 returned. Of my little party of 25 Norfolks, 12 died.'

Over 37 percent of Britons in H Force didn't make it. In all,

22.9 percent of the British assigned to the Railway died, 21.5 percent of Australians, 19.1 percent of the Americans, and 15.5 percent of the Dutch and Dutch Indonesians. The death rates for Malays, Burmese and Javanese were astonishing: 56, 44 and 38.7 percent, respectively.

'For myself I had my reward in the friendships I had made. In spite of everything I shall always have many pleasant memories of the sacrifices, goodwill and considerate treatment and kindness everyone showed to each other. I have never met such a spirit before of self- sacrifice and service.

The railway and its horrors were behind us.'

Or were they?

17 WAKING UP IN PARADISE

**Sime Road Camp, Singapore,
19 December 1943–5 May 1944**

This camp was a British-built hutted one with concrete floors or raised wooden platforms, verandas, modern sanitation, while all the buildings were connected with each other by concrete or tarmac paths and roads. It had originally been built as the HQ for Malaya Command and, although badly knocked about in the Battle of Singapore, was still serviceable, although many of the original huts no longer existed.'

It faced the Royal Singapore Golf Club, founded in 1891, the manicured greens overrun by the advancing Japanese, heading for Percival's headquarters. He vacated Sime Road on 11 February 1942, having woken to the sound of machine-guns across the course, commanding the rest of the war from Fort Canning above Singapore city. By 14 February Japanese tanks were operating along this road. Japanese units occupied the club house, some officers practicing their golf swings the day before taking Singapore.

'Electric light was available and a moderately good water supply from the town mains. It was well laid out with grass and fruit trees and it was like waking up in Paradise.'

Around this time, the song *Oh What a Beautiful Morning* was released to much acclaim and would've provided the perfect soundtrack for Pilk and the 1800 survivors of F and H Forces interned here had they been able to pick it up on their 'canaries'.

'Previous parties, some of whom had arrived over a month before, had got it into working order with an excellent canteen and café. The latter run by the Dutch, cakes, fruit and coffee were cheap and good and, for about fifty cents, one could have quite a good tuck in.

For three days our party were kept in strict quarantine till the Japs were satisfied that we were free from contagious diseases. How they arrived at this decision we never learnt.

I found Monty and Terry Daykin, who had come down by a

previous train, also in quarantine. ----- [name deleted by Pilk] tried to make out he'd never received the Battalion records I handed over when leaving Singapore. Fortunately I found that CQMS Bullen had still got the full records in his keeping and it appeared had refused to hand them over.

I was horrified and disgusted to learn that as soon as I had gone, ----- [name omitted by Pilk] sold all this tinned food to the officers left behind at the cost price, himself taking a goodly share. An examination of the cash books showed that no money was required for the men until Christmas when the prices of the tinned food sold were being quoted at 500 and 600 times higher than those given by the officers. In fact, through premature sales, officers had benefited themselves considerably. His excuse to me that it would have been difficult to distribute anyhow does not hold water. The men were literally robbed of their profits. This was one of the worst cases of the petty abuse of authority that I came across.

On 22 December we were told that a special area had been set aside for officers and we – Monty, Terry, John Packard and myself – moved into a nice room in a row of huts that had a veranda, back and front, a bathroom and a pull-the-plug. Here we settled in comfortably. The others all got hold of bug-ridden beds but I preferred a nice clean door raised on wooden blocks.

All through my time here I stuck to my door bed, much to Monty's annoyance who kept producing the most awful contraptions he had rummaged off scrap iron dumps for me.

We always had bugs, bugs, and bugs. At night, inside one's mosquito nets just as one was going to sleep, that sickly almondy smell would tickle my nostrils and I knew that my bedfellows were on their way. Unless dead tired, sleep would then be impossible till they desisted from their activities early in the morning. I always was dead tired, so generally slept well. Once a week we went through our beds and bedding with a tooth comb, searching the springs and crevices and spearing these unasked guests with a sharp instrument.

Probably the most nerve-racking experience at night is to hear bugs dropping from the ceiling on to the floor around you and the roof of your mosquito net with a dull thud. This invariably happened when one changed quarters. I suppose they all had to come out to see who the new bloke was.

I reacted with horror and disgust when I first found lice on myself and found them breeding in the string round my neck to which my identity discs were attached. Lice and scabies were constant companions till our disinfector was at work. The order was that each room of four officers was to be disinfected at one time; bedding, clothes and everything except leather and books. To do this we trudged the half mile to the disinfector with our worldly possessions hung around us. We put everything in to the big drum to be steamed for half an hour while we danced around it, stark naked, for three- quarters of an hour. It did the trick and we had no more lice. The disinfector, however, had no effect on a scorpion found later walking out of Monty's hat. Spiders and bed bugs all seemed to enjoy their Turkish bath, but I have never had lice again.

From the start we were treated as a convalescent camp, that is to say that we never had to supply working parties for Singapore, but merely had to administer our own camp and get gardens going, cut our firewood and generally look after ourselves.

They started camp workshops where one could get literally everything done from the repairing of a watch or soling of boots, to the construction of complicated kitchen heaters, cooking utensils and excellent clothing.

Before long a daily news service was run. Several searches by the "Gestapo" proved fruitless as we were always warned by our guards when they were coming so everything was nicely tucked away. By Christmas we were all comfortably installed, a reasonably good cookhouse operating under the Dutch, and batmen once more in attendance. The Jap routine at this camp, which never varied, was reveille 7:45, roll call 8:15, Breakfast 9 am, Lunch 1 pm, Tea 4 pm, Dinner 6 pm, roll call 7:15 pm, Lights out 10:30 pm. Christmas and Easter and St George's Day were treated as special festivals when we had a real blow out. Our room had curried prawns, fried pork, spaghetti, tomato chutney and sweet potatoes as a supper dish, Monty building a magnificent little stove behind our room. I stole the firewood, Daykin and Packard skinned the prawns and potatoes and generally got the food ready, while Slap, our batman, and Monty did the cooking. It was a royal party, so much so that we lay in a coma for some hours after.'

Having just spent his third Christmas away from home, and still wondering if he'd be home for the next one, Pilk's hopes were dashed once more: on 27 December 1943 Japanese PM General Tojo Hideki, in a speech to the Diet, warned that 'the real war is just beginning'.

'The Japanese were very worried about us, as it had even sunk into their dense craniums that they had behaved unlike any civilised power in their treatment of us. The result was shown in many little ways. Medical supplies were poured in, efforts made to supply us with books and reading matter, and for this, a very generous and satisfactory ration scale introduced. Rice 13 oz, beans 3 oz, salt, fresh or dried fish, sugar 2 1⁄2 oz per week, tea, vegetables (sometimes green bananas but generally quite good), meat 4 oz per week, palm oil or some other cooking fat. To this must be added our purchases from the camp messing fund of *towgay* beans, palm oil and whitebait, sweet potatoes and — for the hospital — sugar, tea and eggs.

Although we were considered sick, we were actually on the heavy duty scale of rations for POWs. Again, if it had not been possible to purchase extra supplies, the diet as allowed by the Japanese was, for a European, one of starvation.

At home to supplement our meals we had our little gardens where we grew potato tops, onions, Chinese turnips, etc. with which we made a thick stew adding to it towgay, pieces of dried pork and other ingredients that we could buy at the canteen. Roast sweet potatoes were also a favourite and could be bought for 30 cents a time. We also purchased sugar for breakfast and coconuts which we grated and mixed with our porridge. One way or another we looked after ourselves.

In the officers messes we subscribed $10 per head per month with which we purchased still more palm oil, dried fish, *blacan* sugar, coffee, whitebait and sweet potatoes so that, once we had settled down — although we were in practice feeding ourselves — we were feeding well. The normal menu was then breakfast – rice pap, palm oil and sugar; lunch – 3⁄4 pint rice, a good stew, soya bean either fried in palm oil or soaked and baked, and a rissole (doover); tea – small rice biscuit with home-made jam; dinner – good thick green vegetable stew, possibly three tartlets, full of meat, jam or some other concoction, half pint rice, 1/4 pint beans, gravy and possibly a rice cup sweet. Sometimes we had soya bean bread or cake. We had enough if not plenty.

To supplement this diet a café was built and run by the Dutch, known as the Flying Dutchman. Here were seats, electric lights, flowers and creepers, a proper counter and cash desk and such normal fittings as cups made out of milk tins and plates out of pieces of tin. Here one could buy little native cakes, jam puffs, coconut jelly, sweet coffee, cold tea and savoury sandwiches. For lunch they always made a *sambal* of fried ground nuts, chillies and other ingredients to be mixed with one's rice. We had many very pleasant little parties down here.

The walls were panelled with big asbestos sheets which were decorated by Searle, the *Lilliput* and *Gentlemen Only* artist, and Sprod of the *Sydney Herald* [*sic*]. Their drawings were clever and amusing.'

George Sprod lied about his age to sign up, and became a gunner in the 2/15th Field Regiment, Australian Artillery. He and Searle teamed up to produce the magazine *The Exile* for fellow POWs. Post-war he moved to England, teaming up with Searle again on the *News Chronicle*, but became most noted for his cartoons in *Punch*.

'When you are a prisoner of war and too sick to be able to do anything except the lightest work, time lies very heavily on your hands. When, added to this, you have been forbidden to read for more than an hour a day because your eyes are suffering from a vitamin deficiency, the desire to do anything rather than nothing is overwhelming.

A small edition of H.G. Wells' *History of the World* came my way and was my constant companion. [This was the same sodden title which Pilk had discarded en route to the Railway.] Our pay throughout this period was $52/50 per month less 10 percent to the CMF. From now onwards we also subscribed to a fund for the women, children and sick internees in Changi Gaol. Each unit ran its own little fund to help out the sick in hospital with eggs ($1 each), bananas, pineapples and homemade jam. A friend of mine started a still and occasionally we bought, very secretly, a bottle of his gin for $8 and had a celebration.

Tobacco of quite a good variety could be bought for 30 cents an ounce, while Wog Weed only cost $1 a pound. A mixture of the two made quite a good pipe smoke. Cheroots of excellent quality were cheap and good.'

By now, things were getting more desperate on the Japanese home front. They had a real battle on their hands with Chiang Kai-shek in southern China, and the surging US Fleet in the Pacific had just launched the world's greatest warship, the *Missouri*. Tellingly, US battleships shelled the Japanese homeland for first time at Kuril Islands. The Japanese air force was losing 40, 000 fighter planes a year, yet their factories could only churn out 20, 000 replacements.

America was grooming itself for a coming-of-age party of sorts. Roosevelt presented a hawkish US$99.7 billion war budget in 1944, possibly fuelled by the January news that 5200 US POWs had been tortured and starved by the Japanese in the Philippines.

On 4 March 1944 Japanese students began being called up for military service ahead of the Japanese thrust into India. Despite Tojo becoming Chief of Staff of the Japanese Army just weeks earlier, many saw this as Emperor Hirohito's very own offensive initiative. But the Japanese, who had acquitted themselves admirably in the jungles and mountains of Malaya, struggled with logistics from Burma to India – mainly resupply issues given that they did not have much in the way of bombers, let alone cargo planes – and were on the receiving end of their greatest land defeat ever.

(During their push on Arakan, the Japanese entered a British field hospital where they bayoneted the wounded in their beds and massacred the British-Indian medical team. It was the third such atrocity, following the Alex and St Stephen's College in Hong Kong.)

To further upset the Japanese game plan, their Co-Prosperity Sphere was now suffering from commodity and labour shortages, and inflation raged unchecked.

'In March an American Red Cross ship came in laden with food parcels; from accounts since received, probably 10 or 12 per head. After a long delay the Japs allowed us one parcel between eight.

These parcels, however, with their tins of salmon, sardines, meat roll, jam, sugar, chocolate, butter, cheese and cigarettes made for ten days a good change in our menu. All the articles that could be economically used in the kitchen we handed over and they made excellent dishes with them. Other items such as cheese, chocolate and cigarettes and soap were divided out with meticulous care. I shaved with the excellent soap for nine months.

Initially, the Japs produced thread, old shorts and sewing machines and our tailors made a splendid job of this second rate material. Latterly, as the Japs refused to supply sewing cotton, we manufactured our own.

My clothes by now, in particular my shorts, had almost completely fallen to bits but, as none of the new clothing supplied would anywhere near fit me, no effort was made to get me some. I decided to stop patching with the result that, within a few days, all the Colonels were on the track of the QM who produced me clothing made in the camp workshops in double-quick time. I was delighted with my clothes but more so with my moral victory, and spent many happy hours explaining my methods to others, while the QM — who lived on the other side of our thin partition — listened in uncomfortable silence.

Everyone was expected to do some work for three or four hours a day again. Most of the officers gardened but Jacko refused to let me out in the sun as the bright light aggravated my eyes and made them inflamed. I therefore helped in the officers' kitchen peeling vegetables we had robbed the pigs of, grinding pepper in an antiquated stone mill, chopping up bananas, etc.

For the rest of the time I was my own master and could at last sit back and fulfill an itch I had for months to write.

During these months I amused myself by rewriting the long stories I had written in other camps. The writing of stories did at any rate pass countless weary hours and enabled me to live in a fairy world that had no connection with my prison. Great rolls of barbed wire imprisoned us only too well.'

These stories included *Bungalow Days*, *Ornot's Ornithology*, and *Silver: Paul's Own Story*, all full-length manuscripts written longhand. How Pilk must have reflected on his change in fortune when reliving his enjoyable rubber planting days in Malaya, captured so evocatively and fondly in *Bungalow Days* —

now, in 1944, even the wealthy throughout Malaya were down to one rice meal a day. *Ubi kayu* (tapioca), yams, and sweet potatoes became the substitute staples. Malnutrition stunted the growth and development of war-time youngsters.

The Japanese encouraged the people to return to the land and plant more 'victory crops' such as tapioca. But many from Singapore moved by their own choice to smaller towns where there was less chance of crossing the Japanese, and it was more peaceful and safe. Pilk's mind was also very much on the welfare of Phyllis and Paul.

'I had the idea of making bed-time stories for Paul, each of which would perhaps impress upon him some small point and even arouse curiosity by details which might bring him an added interest in life and make him appreciate the world around him. When the stories were written I had not seen Paul and could only imagine what would interest him.

The manuscript for *Ornot's Ornithology* was preserved for me by our British Japanese interpreter being able to convince the Japanese commandant to put the official Jap "censor" stamp on the manuscript. It was at one time taken by a Jap "Gestapo" search party as they were convinced it was an escape plan.'

For Pilk's sake he was lucky, or he might have found himself in cell number five, which the Kempeitai used as a torture cell. This measured just 5' x 8', and was often occupied by six people. Giving guards nicknames was a popular pastime, and among those at Sime Road were Okay Charlie and Puss in Boots.

'Paper was always a difficulty and was bought at around ten cents a sheet. No such luxuries as a chair or table existed in my quarters so the whole thing had to be written on a board propped up on my knees on my bed.

I am indebted to many friends who not only gave me what plain paper they had carried through the Thai jungles, but for the way in which many of them assisted me to purloin paper from the Japanese when their backs were turned.

The illustrations for *Bungalow Days* done by Ronald Searle were all executed under similar conditions and their colouring was governed by what coloured pencils or paints he could get his hands on.

Concerts, theatres and lectures were started. At first we had an open air theatre built in a natural amphitheatre. Later this was changed to a hutted one of great ingenuity with benches, pay-box, drop scene, bills and even a fire curtain. Entrance fee 10 cents. Here the theatrical party produced a homemade and written pantomime entitled 'Cinderella and the Magic Soya Bean'. Very clever, excellent costumes and very funny. Mostly it was a skit on the Japs. They enjoyed it. Other plays produced were *Man of Destiny* by Bernard Shaw, *Rope,* and *Bird in Hand* with numerous productions of the concert hall variety.

Every Thursday the Colonels had an impromptu talk in their hut by some officer to which they invited their friends. I was a constant attendant and learnt a great deal of the outside world from these discussions. The evening always finished with coffee and a bun supplied by them. They were most enjoyable.

There were very few deaths in this camp. Nearly everyone made a good recovery. As regards myself, once over my jaundice I suffered no major illness. The jaundice undoubtedly killed the malaria! At the slightest indication of trouble or inflammation in my eyes I was put on that precious substance, Marmite, and rice polishings by Dr Dixon, who, as a good friend of Jacko's, took a very kind interest in me. Beri-beri and other minor deficiency diseases I never shook off, but lying or sitting on my bed all day rested my legs and I was not unduly bothered.'

It would have been comforting for Pilk to learn that on 10 April, quinine — used to ward off malaria — was first synthesised by Robert Woodward and William Doering at Harvard University. Advances in tropical medicines such as this saw Allied forces' disease rates drop remarkably but was out of reach of the POWs.

A couple of weeks later, the Allies launched a major attack in New Guinea. Inside the camps, the illegal wirelesses also picked up the vanquishing of Italy, the Russians pushing towards Berlin. The local rumour mills also had it that Allied armies were poised to strike back in Burma. That the tide was turning was palpable.

'In April a new Jap General, Major General Shito, took over administration of POW and Internee camps. Horrified at the conditions in which the men, women and children were living in Changi Gaol, he ordered a switch over — they to come to Sime Road and we to go to the gaol.

A pig and poultry farm was started but by the time they were getting under way we had to move, so we presented them to the incoming internee women and children.'

Three-and-a-half thousand civilian prisoners from Changi were moved to this former Combined Army and Air Force Operational HQ camp at Sime Road. No one was excused work parties, even ten-year-old children laboured in the surrounding fields for four hours a day.

A guard once beat up 22 internees in a drunken rage. By the end of war 4500 civilians would be held here, representing twenty enemy countries and ten neutrals.

'A great deal of the credit is due for our humane treatment over the last months to Lieutenant, later Captain, Takahashi of the Japanese Army Engineers who administered our camp and I still believe he did all in his power to help us, although he was thwarted in many cases by his superiors.

A lover of animals he was always accompanied by a pleasant and mischievous monkey, while his house was full of really good Siamese cats.

While here I was able to study at my leisure Wells' *Outline of History*, which fascinated me. I also read Walpole's *Herries* saga, with its fascinating description of Cumberland.'

By 3 May meat rationing had ended in the USA — a positive sign. Yet, paradoxically, the better the war went for the Allies, the worse conditions got for POWs in South-East Asia.

'On 5 May we filed into lorries with our beds, furniture, and anything we cared to take, and said goodbye to Sime Road.

A really happy existence terminated.'

‘The Sime Road camp (above) Pilk likened to 'waking up in Paradise.' (David Tompkins and Alex Turnbull Library) 'Never have I known a building that cast such utter gloom over me,' said Pilk of Changi Gaol. (AWM 117645)

18 DOOM AND GLOOM

Changi Gaol, Singapore, 5–9 May 1944

'Before leaving Sime Road, the internees started to arrive. By shouting across the no-man's land between us we were able to get a few messages to them. The women and children looked fit and smart. One small boy of about seven amused us by going up to the Sikh guard, shaking his fist at him, and calling him a "*Babi*"" [Malay for pig]. The Sikh looked very uncomfortable and sheepish.

By good luck I picked up an almost brand new Indian *charpoy* (string bed) which, as it could be taken to pieces and so could be debugged, I commandeered. It was with me up to the day we left for England. I also got four pounds of tea that was left behind which proved in the days to come an absolute godsend.

We moved off in lorries, about 25 to a truck. Not too bad as the Japs usually considered forty the minimum. On the way to Changi we saw a Jap knocked down by a lorry. The utter casualness with which he was treated by his own people was typical of the Jap race. Having looked at him curiously for some ten minutes, they stopped a passing car, hauled him in and sent him off. His cries of agony (he had a broken arm and leg) met with derisive remarks.

Changi Gaol, known to the Japs as "The Big House", is the most modern prison in the East, being modelled on the plans of the American gangster prison at Chicago.'

Changi Gaol was actually modelled on Sing Sing in New York State, a multi-level 'stacked' design, and was the last prison built by the British in Singapore. When it was officially opened in January 1937 it was intended to house 450 prisoners. As many as 3500 civilians were held here by the Japanese in 1942.

'Never have I known a building that cast such utter gloom over me. Surrounded by two thirty feet high walls, one could see nothing of the outside world. Everywhere were concrete walls, steel bars, expanded metal gratings and iron doors. Where there were windows, as in the hospital block, they looked out on to high walls and were so closely barred that one couldn't get one's little finger through. The cells, designed for one Asiatic prisoner, had contained three women or children, and now contained four Europeans. They were in three tiers. Each cell had a concrete sleeping block in the centre, a native type water-flushed WC and a tiny window so high up that practically no light could enter. I and twenty other officers were put on the top storey of the hospital block, the best place in the prison. Here we put up our beds and made ourselves comparatively comfortable. Terry Daykin accompanied me. Monty and John Packard with four others were quartered in one of the life sentence cells, while a lot of others made their home in a grim cell marked 'Flogging Room'.

The only good thing in the gaol was the kitchens, very modern, steam heated, and capable of turning out good food, cleanly prepared. No frying, however, could be done as all gaol cooking was evidently boiling rice and stews. A good bakery was available but at that time no wood fuel to heat it.

Towering over the gaol was a huge clock tower and sentry post. The clock struck the hours and half hours to the chimes of Truro Cathedral. The sound of those bells chiming away the hours of one's life was almost unendurable.

Electric light and water were available up to 10 pm, after which they were switched off. We had been warned that the gaol was fitted with microphones so, to the depressing atmosphere, was added the fear of giving away secrets.

Our food was one pint of pap for breakfast and one pint of tea. Lunch – 3⁄4 pint rice, 1⁄2 pint green leaf stew, 3⁄4 pint tea. Dinner at 6:30 – one pint rice, one pint thick meat stew and pint of tea.

Fortunately we were able to supplement this diet with little bits of fruit and luxuries we had brought from Sime Road.

I, not available for outside work, took over the administration of my block and also the very onerous job of drawing the meals for 400 officers and men, and distributing this fairly.

At first I had a great deal of trouble in getting our full share but a really good row put this right. It wouldn't have been so difficult to dish out if everyone had been in the same block but they were spread all over the gaol. However I seemed to satisfy them all and before we left had got it down to a fine art.

At nights the oppressive closed atmosphere gave me a sort of claustrophobia, making sleep almost impossible. This, however, I was able to make up during the daytime. On the morning of 9 May we were told that all officers were to leave the gaol and go to Selarang for the time being.

At three o'clock trailers drawn by officers from Changi arrived to take our kit over. Nothing loath, we piled it on and set off for the two and a half mile walk.

Although it was one of the hottest days I have experienced, I have seldom felt such relief at leaving a place. Those few days had been a mental hell. Much to our delight, our trailer was drawn by a gang composed mostly of field officers who we encouraged up the hills, when they faltered, with appropriate remarks.

They, however, were delighted to see us back and we had a merry journey.'

19 LIVING LIKE FIGHTING COCKS

Selarang Barracks, Singapore, 9–30 May 1944

In May 1943 when the Changi Camp had been drained dry of every fit prisoner for the Railway, the Japs had closed down all the other areas, concentrating everyone in the old Australian area of Selarang. Recently, to accommodate survivors of F Force, some 3000 odd, they had again allowed the gardens and woods area to be reopened as a convalescent hospital.

Now orders had been given that everyone was to be housed in Changi Gaol or its vicinity by the end of May, with the exception of 2000 of the chronic hospital cases who were to be sent to an old British hutted camp near the Causeway called Kranji.

Kranji, on the north coast just west of the Causeway, was a military camp area before 1939 and, at the time of invasion, home to a large ammunition magazine. After the capitulation, the Japanese established a POW camp there.

'On arrival at Selarang the Australian officers were taken off to their units but the British officers were housed in a row of buildings similar to Quadrant Road. Guided by my love of privacy I made for the scullery at the back of the end house. This I shared with Daykin, Titmarsh and Moor. A tiny room in which we were rather crowded, but being out the back of the house with access to us only right round the back – or through a room in which some twenty officers lived – assured us of being undisturbed. Monty wanted me to move into more palatial quarters but I never did.

Here we lived till 30 May in comparative comfort. Our application for an electric light bulb was refused so we helped ourselves to one of the many that burned in every inhabited house. The water supply was good with shower baths outside our scullery door.

For the first three days we were fed by an established kitchen but we quickly found that not only was our palm oil and other stocks disappearing into other mouths, but that there were unaccountable men taken in in apportioning our rations, always to our disadvantage. We therefore started our own kitchen and from then onwards lived like fighting cocks.

The standard of cooking by now was excellent, many clever and ingenious savoury dishes being prepared. A small meat issue was still being made and, as well as that, we got a little pork from the pigs we had subscribed for two years ago and which were now being slaughtered as there was no room for them in the Changi Gaol area. Needless to say the Japs, unasked, helped themselves to half our stock.

The gardens had been kept up so we got masses of vegetables. In addition we were able to buy *towgay* beans, coconuts, *blacan*, etc. from the canteen as well as cigars and quite a good pipe tobacco. By now prices were rising and commodities getting scarce. Coconuts that had been two cents were now 30 and everything else in proportion.

To accommodate everyone in the area allotted at the prison it was necessary to transfer every wooden hut and re-erect it. Every piece of piping, electric light wire, concrete slabs and flooring, kitchen ovens, beds, furniture, wash basins and everything necessary to house 12, 000 men had to be transplanted by convoys of trailers.

Almost immediately on our arrival we were medically examined and put into various categories for work. I was placed in the lowest category, PUB (Permanently Unfit B but able to do light duty indoors) and this category, in spite of six monthly examinations, was never changed. In fact I was excused all duties but, from now onwards, I continued giving a hand in the kitchen for an hour a day with the vegetable preparation.

In the meantime every fit man was taken by the Japs for work on the aerodrome so that a great deal of this haulage and heavy work fell on the officers who, twice a day, pulled the laden trailers to the gaol, unloaded them and brought them back. In the circumstances it was not surprising that every officer who could had to help.'

The new aerodrome was located between Changi Gaol and Roberts Barracks.

'While here I was examined by the oculist, Colonel Rose,

who took the greatest trouble. Eventually he pronounced my eyes weak but cured, and gave me permission to read as much as I liked in daylight provided I stopped as soon as they got tired. He also gave me a prescription for specs but, as no one could get them for me, I am still waiting. The surgical specialist, Colonel Taylor, also examined me, much to my embarrassment, in the middle of the parade ground. I think if he could have got X-ray photos then he would have liked to operate on my shoulder, but the apparatus was by now so weak that the results were unsatisfactory.

Shortly after my arrival I saw Mike Louth, Hughes, Shorter, and hosts of old friends, all of whom were extremely kind to me, giving me odds and ends of their belongings to make good what I had lost in Thailand, asking me to their private concerts, to coffee and buns, and generally they made life most pleasant for me.

The barracks and married quarters at Selarang were spread over a huge area so one could walk for quite a distance. On the occasion of the unveiling of the memorial tablet in the cemetery to Major General Beckwith-Smith who had died as a POW in Formosa, I went down with a representative part of the 18 Divisional units. [Beckworth-Smith, Commander 18 Division, died of heart failure brought on by diphtheria on 11 November en route to Formosa.]

The cemetery by now had made great strides, it was grassed, many of the trees I had helped select were quite big, and the hibiscus hedges almost fully grown. From there one could see the Padang and Coconut Grove areas where we had been so happy. It was now all cleared and a huge aerodrome had taken its place, the work being done mainly by POWs but under much better conditions, shorter working hours and good food that had been lacking in Thailand. Furthermore, the Japs here would not allow their engineers to beat up prisoners.'

By this stage of the war, Hitler had given permission for a full German withdrawal from the USSR, and MacArthur was steaming up from Australia, having reached Biak Island, New Guinea, 900 miles south of the Philippines.

'On 30 May at 2 pm we had our last meal at Selarang, loaded a trailer, and set off for Changi Gaol once more.

I have never seen any of that or the Quadrant Road area again.'

20 THE TURNING POINT

Officers' Area, Changi Gaol, Singapore, 30 May 1944–10 August 1945

'The notorious Changi Gaol Camp was laid out and planned by our own REs. Nothing could be more compact and serviceable. The area allotted to us by the Japs was the prison itself, and round it outside the walls a space some 400 yards wide on two sides of the gaol. On the third side ran the main Changi–Singapore road while on the fourth side were housed the Jap guard and our hospital.

The officers' area took up one section. The other area by the wall housed some 3000 other ranks in long *atap* huts about 300 feet long. This area had its own kitchens, showers, etc. and was run independently to the gaol. There was, however, no water-born sewerage there.

The officers' hospital was housed in the last four houses of the European warders' section, the remainder of these buildings, some 24 in number, accommodating the Jap guard.

Into the gaol the Japs had crowded 6000 POWs, but a special area outside the prison walls was allocated to the 1200 officers that were in the Singapore camps. The officers' camp was built of long wooden huts arranged in rows on either side of the brick and concrete coolie lines originally occupied by the prison warders. Each coolie line had a room 12' x 14', a small veranda 6' x 19', a tiny kitchen 8' x 6', a small bathroom and a native water-flushed WC.

Obviously these were the pick of the accommodation and, thanks to my seniority, I was allocated to one of the rooms with four others. At first I occupied the corner of the room by the door when I was comparatively cool and comfortable.

By the time our beds were in, a table and chairs on the veranda, and our little bookcase and tables – without which life seemed unendurable – were installed, there wasn't much room, but we were infinitely better off than those in the wooden huts who, 40 or 50 to a hut, had only sufficient space for their bed and chair.

Our beds varied from concrete floors with a mat over them, to wooden plank beds and even vi-spring mattresses. I preferred the native spring and wooden bed known as *sharpoy*, as they were cool, not too springy, and easy to take to pieces and reassemble in a new place.

I again took to bridge playing nearly every evening with Mike Louth as a partner. We played for love and had many a good laugh. The evenings passed quickly, some badly needed exercise for one's brain was procured and, by combining my old auction bridge methods with the new contract ones, we put all Mr Culbertson's satellites in the shade. Without bridge in the evening the nights would have dragged endlessly.

Behind our quarters was our cookhouse, built by ourselves to accommodate the 180 officers and OR in our group. The smoke and noise from this cookhouse was an eternal nuisance but if we complained we could only be moved into a hut which was worse. The cooks, however, were a good lot and did everything they could not to disturb us unduly.

At first in our kitchen we had two batmen but later it was turned into a storeroom and kitchen officer's office. This brought additional flies into our room and meant that to read in the afternoon in comfort one had to do so under one's mosquito net, the heat being preferable to the flies.

Swotting parties had to be detailed to keep them down in the kitchen. Their tickling feet were far more aggravating than the sting of a mosquito or the smell of a "bed bug". Some units even had a Fly Hunt with a Master, whips and hounds, and worked to very strict rules.

The Japs loathed bed bugs and, at the first itch, he would strip to the skin and search his clothing. Knowing this, our men collected all their bed bugs in tins and — when ordered to clean the Japanese barracks and huts — scattered them ad lib over the whole place. It was not surprising that they changed their quarters frequently!

The hospital was always appallingly overcrowded but no representation made to the Japs ever produced any result. The area, owing to its crescent shape, was known as Half Moon Street.

The doctors and orderlies lived in wooden huts on the slope of the hill in front of Half Moon Street. Their area was known as the Quackery.

Behind Half Moon Street was a row of houses that had been occupied by the head Asiatic warders. Here was located the operating theatre, the surgical wards, the various clinics, eye, nose, medical specialists, etc. In fact this was the kernel of the hospital.

Around it were built wooden huts to accommodate the less serious surgical cases. Beyond them again were constructed a row of *atap* huts, some twelve in number. Each hut was 300 feet long and capable of holding 100 patients in comfort. They were allotted to dysentery, malaria, skin diseases and so on.

Flanking these buildings were the hospital kitchens, the mortuary, isolation ward for infectious cases and a TB ward, the latter two being wired off from the rest of the hospital.

In this hospital area were accommodated some 2000 officers and patients. It was well run and well administered and the comparatively light death rate for this period speaks very highly for the work of the RAMC.

We had for the hospital a small herd of goats ruled over by a charming billy that I fed with grass each evening and which would stand by the fence watching for me. When I was about 100 yards away he would recognise me with delighted bleats. Pigeon lofts and rabbit farms were instituted. Beehives gave their lucky owners honey.

On the outskirts of the camp were three areas allotted to poultry, very strictly supervised, carefully laid out and spotlessly clean. The threat that all poultry would be slaughtered if any fly breeding was discovered assured this.

Inside the wire, every inch of ground was brought under cultivation of vegetables, special gardening squads being detailed for this purpose. Each group had its own area to cultivate, the vegetables so obtained being in excess of those supplied by the rations. Outside the perimeter barbed wire fence were large Jap-supervised gardens, worked by our men and officers, on which we eventually practically lived.

All urine as well as the sludge from the septic tanks was collected to manure the gardens in special carts. Salt water was collected in water trucks from the sea for cooking purposes. Firewood was felled and carted by our own people. A grass extract factory supplied vitamins to a certain extent. Palm trees were tapped for toddy and half a pint a day allowed to the very worst beri-beri cases. If we couldn't produce an article, we couldn't get it.

As long as we could smoke we didn't feel so hungry, but, as the food got shorter, so the craving for a smoke increased. As supplies of food became harder to come by, so tobacco became almost impossible.

Cigarette ends were hoarded from the first days. Cigar and cheroot butts were washed, dried and re-cut into tobacco. From May 1944 onwards tobacco became an ever more serious problem and, before long, everyone had their own recipes for making mixtures out of cherry and brinjal leaves, cigar butts and cigarette ends, Wog's Weed (like dried rank hair) and other stuff. Some people would boil everything up with a little *gula malacca* and put it through complicated processes of drying and pressing. Others just dried the leaves and chopped them up. Whatever the results were they were generally just smokable.

A good smoke was a joy and on tobacco and food went all my money.

As regards food, when we first went to the prison we were given 16 oz of rice a day, two oz of soya beans, two oz of meat per week, eight oz dried fish, one lb of sweet potatoes or tapioca root and 1/2 lb of green vegetable (either *kang kong* or Chinese cabbage), a small amount of palm oil, a little sugar, a little salt and so much pepper we didn't know what to do with it. To this must be added a liberal issue of *towgay* beans, whitebait and palm oil from the camp messing funds. Our mess subscriptions brought us coconuts, *blacan,* dried prawns, curry powder, sugar and other small items that we used to make the meals more attractive. At first we lived very well and no one went hungry.

Matches very soon gave out so we had recourse to cigarette lighters filled with petrol siphoned from Jap lorries. Long lengths of smouldering string. Little lamps of palm oil and until finally every quarter was fitted up with home-made electric cigarette lighters.

Thieving of petrol they could never stop. On one great occasion a Jap general gathered all our transport drivers together and gave them a lecture on sabotage and the stealing of petrol and oil, and took great pains to explain that anyone found doing so would be shot. He got into his car, drove half a mile down the road only to find that while he had been lecturing our men had drained his tank dry!

Bicycle chains and tyres fetched a fabulous price, and any Jap so foolish to leave his machine unattended was certain to lose the chain, if not the tyres. Incidentally, a bicycle chain fitted nicely into an army water bottle.

We heard that an Australian POW camp in Singapore had induced the Japs to buy a hut they had stolen to house the lorry that the Japs used, and also to house cement. The Australians sold piecemeal the bags of cement, then the hut disappeared, and finally they flogged the lorry to a Chinaman!

There was a godown at Singapore docks full of plum puddings. For a long time our men had helped themselves to one apiece with the knowledge of the Japs till one day one man took two. The resulting beating up of all and sundry was one of the worst on record.

When a German raider was in Singapore dock being refuelled the German sentry saw a Jap beat up one of our men. He and his companions promptly dropped their rifles, beat up the Jap and threw them into the dock. Diplomatic relations are said to have been extremely badly strained by this incident.

We were never without news. The "Gestapo" threats of shooting and mass punishment were no deterrent. The news, called the canary, came through. Every day at 6 o'clock the news bulletin was given to us in detail by word of mouth. We were always better informed than even the Jap HQ!

As the Japs got more and more inquisitive so more and more ingenious became our operators till at last a very secret set was constructed that could be fitted into a wooden bed frame.'

June 1944 was a major turning point. Developments were eagerly lapped up by inmates at Changi who covertly listened in on their 'canaries'. First, the Allied forces landed in Normandy. Then on 6 June, American B29s dropped their first bombs on mainland Japan since the Doolittle raid of 1942. Up-country even the Thai–Burma Railway was now under Allied aerial bombardment, adding to the danger of those who remained behind to maintain it.

By 15 June, bombing of Japanese home islands from Allied land- bases in China had started. The Japanese intentions on India were also thwarted, setting the stage for the clearing of Burma by June 1945. Significantly, this gave the Allies a toe-hold back onto the sub-continent, restoring both prestige and morale. The Japanese were now defending. Although they were mustering the same troop numbers as in their initial 1941–42 offensive, this time they were without the initiative and element of surprise. Within a week, they'd lost 400 planes and three carriers in the battle of the Philippine Sea.

Through all of this, canine-loving Pilk finished writing *Michael*, his story of a dog.

'In June 1944 the meat ration was cut and the fish ration halved. No one was ever late for meals. As food got shorter and shorter, so the measuring out became stricter and stricter. Spoons and ladles of all sizes and calibres were made and these would be filled and levelled off with a scraper to ensure fair play.

Leggees [a corruption of *lagi*, the Malay word for more] were instituted at the very start. Leggee lists were posted and anything left over was shared out in turn. Woe betides any man who tried to get in on a leggee queue out of his turn.

Food was always predominantly on our minds and everyone had a few recipes that he treasured for the future. Some people had whole volumes running into the thousands of recipes of all kinds; others contented themselves with some well-worn ones.

Some collections were extremely practical and confined themselves to dishes made with rice and local foods. Others soared into realms not likely to be achieved under our present rationing system for many years.

After food came our homes. Everyone designed and furnished the house of their dreams. Some extended to private theatres and war memorials. So great was the passion for our future homes that exhibitions of drawings were held; architects made out plans and even accepted commissions.'

Pilk drew elaborate double-storey mansions, with grandiose gateposts.

'These recipes, drawings and jottings were our most

guarded possessions next to our letters from home, and, if our dreams had not yet come true, they kept our minds actively engaged and prevented us dwelling on the miseries of our captivity.

As our morale must always depend on our personal appearance to a certain extent, we were on the whole well turned out. Everyone washed their clothes, darned and stitched and generally did everything to keep up appearances. While razor blades were plentiful we shaved every day but, as they got scarcer, we had to shave twice or three times a week. Later barbers' shops were set up, complete with looted chairs, where skilled barbers with excellent razors made from army dinner knives did their duty.

We all kept our hair short and it was amazing the number of skilled barbers that appeared and who made a good side business at 10 cents a time. Many adopted the Jap habit of having their heads shaved. As I never got nits I never did.'

A Ronald Searle cartoon of Pilk in Changi showed him with towel around waist, smoking his pipe and carrying a cooking pan. True to form, his moustache looks impeccably groomed.

'Very rapidly Changi Prison Camp became an almost self-contained unit except for the production of rice and fish. Metalworking and carpenters' workshops were opened, an electric welding plant manufactured our cooking utensils, watering cans, etc.; a nail factory supplied us with this otherwise unobtainable article; a sawmill produced our timber; cotton thread for repairing clothing, paper for writing and records, tooth powder, soap, artificial teeth, rubber footwear, clothing, some drugs such as magnesia, were all manufactured. One could get anything done from a watch repaired to a book bound.'

Indeed Changi would come to be regarded as the best camp to be in under the Japanese anywhere in the Asia/Pacific theatre. If food, medicine or support of any kind was available to POWs, Changi camp would be the place to find it, earning it the relative sobriquet of 'a resort', 'a country club' and, according to the SEAC All-Services Newspaper of South-East Asia Command, 'the luxury camp'. That the word Changi has become shorthand for atrocities is lazy shorthand stereotyping.

'The great intensity of the way everyone did everything impressed me so much. No matter what the job, it was done with such care and diligence that one would think that one's living depended on it. Captain Clements of the RASC, one of my room companions, sat for hour after hour with my broken watch in front of him while he manufactured a new part for it out of a needle, using home-made tools and files, with a jewellers' magnifying glass glued to his eye for hours at a time.'

Throughout Singapore and Malaya, there was a growing feeling among the populace that the Co-Prosperity Sphere was floundering. Even in Japan a groundswell of anti-war sentiment was growing, reaching boilover with Premier Tojo Hideki dumped as Chief of Staff on 18 July.

'Outside my coolie quarter was a Japanese cherry tree. This I trained into a shady arbour with wires and sticks and eventually we had all our meals in its shade. For months, early every morning, I gathered at least a pint of fruit off it which I mixed with my porridge. Soon also it was the haunt of sparrows, yellow-vented bulbuls, tailor birds, fly catchers and flower pickers, encouraged by crumbs of rice I put out for them. They gave me the greatest pleasure.

Under the tree for a long time lived a delightful dog which was almost a full pedigreed Kerry, called Jeremy, the property of an old RMR officer, Lieutenant Knight, who looked after him with the utmost devotion. Much to our sorrow Jeremy eventually died of beri-beri and pneumonia.

Unable to work outside, my duties here consisted in supervising the officers' kitchen vegetable fatigue which I did every day without a break.

During this period sweat, and being unable to lift my left arm, caused very severe inflammation in the armpit. This was successfully cured by the skin specialist after a long and concentrated treatment of violet rays. Fortunately at this time a new ear specialist appeared who did wonders for me, himself carrying out the treatment, with the final result that all the inflammation has gone and, except for deafness, they are quite cured.'

Come August 1944, the expulsion of the Japanese from Burma was seemingly inevitable. On 10 August, the Americans turned up the heat on Japan, and launched an air raid over the strategic city of Nagasaki.

'By August the Japs had turned out $25, 000, 000 paper currency and the Chinese $50, 000, 000 in counterfeit, with the result that money became so cheap as to be almost worthless. The Japs allowed us to draw an extra $10 per head per officer. The men were paid by the working day and received almost $8 per month, just sufficient for one oz of tobacco or 1 1/2 coconuts.

A large central library was opened for officers from which one could get books in every imaginable subject. To join one had to put in a book. Now that I was allowed to read again, I settled down to read all the classics I had wanted to for years. Undisturbed, I passed many hours away. Camp magazines, most of them quite futile, were started, but it gave an outlet to would-be authors. Classes were held in every known subject and language. I attended the Early History classes by Dr Gray, which were extremely good.

In the officers' area a small theatre of the open-air variety was built where lectures, discussions and plays written by the officers themselves were produced. The Americans produced some very good gangster plays — *Speak Easy* and *Three Men and a Horse*. Others acted here were a revival of *The Dover Road* and *Colonel Fothergill's Party*, the latter absolute trash.

In one of the gaol courtyards a magnificent theatre was built with drop scenes, the latest lighting effects, proper orchestra box, curtain and all scenery: a proper little theatre that seated 700 at a time. Here were acted by a cast drawn from the best talent in the camp, *Autumn Crocus*, *Jupiter Laughs*, *Red Pepper* and *Tonight at 8:30*. Several excellent concerts, an orchestral display with, on Sundays, a symphony orchestra.'

Pilk makes no mention of *Changi Evensong*, which was apparently sung most nights during captivity. Composed by John L. Woods, its first verse runs: 'Around these prison walls which long ago / Became our home beneath the Rising Sun / Darkness has fallen once again, and lo! / Another day is done'.

In August the Jap palm oil ration nearly disappeared altogether and it was unpurchaseable except for medical requirements, as they were using it as a lubricant in their lorries.

September 1944, the momentum continued swinging the Allies' way. The Japanese, no longer able to hold north Burma, were hastily retreating south. Thus encouraged, on 11 September,

Churchill and Roosevelt met in Canada to discuss their strategy against the Axis powers, plus their plans for the post-war world.

On 21 September, a US carrier force struck around Manila. That same day, the *Daily Sketch* carried an article headlined: **'PRISON SHIP RESCUES'**. This told the seemingly unbelievable story of how 159 British and Australian captives were rescued by the US submarine *Pampanito*, after five US submarines (including the *Sealion*) sank the *Rakuyo Maru* with 1300 POWs on board. 'All were exhausted and showed signs of malnutrition,' reported the US Navy.

This was the moment the rest of the horrified world first learnt of the atrocities of the Thai–Burma Railway and the harsh conditions endured by Allied POWs. The silent insulation of Malaya had been ripped asunder with this thunderous revelation. The *Rakuyo Maru* was one of several unmarked 'hell ships' to be sunk, usually by American submarines. In all, 10,800 POWs died at sea (about one-fifth of those transported), only marginally less than died on the entire Railway. That it was by 'friendly fire' made it even more galling.

'By the end of September with the Allied blockade tightening, food prices were soaring and canteen supplies only came in a trickle. For breakfast four of us shared a coconut which we grated in turn and mixed with our pap.'

Even Japanese soldiers were down to a 400-gram daily ration of rice.

'Money problems we always had. Those who had made money on the black market sold it against English cheques. Everyone tried to find some source of supplementing their incomes. Some kept bees and sold honey. Others sold eggs. Some did leather work, typing, and repairs. Others were spivs who contacted the black market. Those like myself who found it difficult to find any means of supplementing one's income had to sell off a shirt or pair of shorts occasionally.'

It wasn't just Pilk and the POWs feeling the pinch of inflation. The entire region was suffering as supply lines got progressively shut down. In October, the Japanese closed their Singapore–Medan shipping route because of British submarines operating in the Malacca Strait. Double the volume of war materials was pouring into China and North Burma during 1944, and the Allies were rebuilding railways and docks in India and Burma. As General Slim's 14 Army was pushing ahead up into north Burma, his Japanese counterparts were recalled to Tokyo.

In the middle of October, a record number of B29s blasted Okayama, further diminishing Tojo's own position and weakening his standing. On 20 October, MacArthur waded ashore on Leyte, Philippines, fulfilling his self-promotional promise of returning.

'In mid-October a bad attack of beri-beri, combined with a deficiency disease in my mouth and throat causing big sores, cracked lips and an ulcerated throat, intervened and I was excused all work till mid February. After that, I only appeared in the kitchen twice a week.

There was no real cure for the mouth or throat deficiency diseases, but I was given palm oil — by now very scarce and only available medicinally — a pleasant strong mouthwash, and a pint of a filthy concoction of grass extract each day. This was made by chopping up grass, soaking it in water to extract the juices, and making the patient imbibe the result. It was filthy but undoubtedly did us good.

On 5 November, Guy Fawkes Day, came the first air-mail. Ninety-eight Flying Fortresses bombed the naval base, to our cheers. We could, when the Jap guard was out of the way, see and hear everything. The raids continued at varying intervals.'

The Flying Fortresses flew down over Bahau (centre of the Malayan peninsula) and took a direct southern bead from there. With these regular raids, the Malayan population felt the end of war was surely near. This feeling would have been heightened on 12 November when US fliers wiped out a Japanese convoy near Leyte, accounting for some 8000 troops, six destroyers and four transports.

Saipan-based B29s even bombed Tokyo in broad daylight, the ultimate military humiliation.

'Every month for a year as the moon rose we said, 'This is the right moon'. As the month passed everyone found reasons in regard to winds and weather why the invasion must at any rate take place next month. Only those ghastly people that styled themselves "Realists" talked of the war and our captivity lasting months and years.'

On 24 November — a full two months after the rescue of the POWs off the *Rakuyo Maru* — acting Australian Prime Minister Frank Forde made a public statement on the treatment of POWs in Singapore and on the Thai-Burma Railway.

In line with the hit song of the day, *Accentuate the Positive*, Christmas 1944 could be viewed as a good one for the Allies. They had now crossed the Irrawaddy, a psychological turning point in retaking Burma. But ahead was the grim and gruelling prospect of fighting all the way down the coastal crescent to Singapore.

'Up to Christmas the Japs kept up the vegetable ration which at least blew us out and made us feel full for an hour or two. Every month we were weighed. At Christmas my weight was 13 stone 10 lbs (peace time 16 stone 7 lbs).'

Coming off the Railway, officers generally could count on weighing a good ten kilograms heavier than other ranks due to lighter work loads and/or more rations.

'Food-wise, our great ambition was a "knife and forker", realised only on such great festivals as Christmas Day. For Christmas we had hoarded little supplies of luxuries, and our supply officers had also a little hoard and were able to make a double issue. For the last time for many months we ate to repletion, and I was still eating bits of my Christmas fare three days later. The cook, working night and day, produced imitation mince pies and Christmas pudding, apple charlotte, imitation mixed grill and I don't know what. For three days we were all happy and everyone was wreathed in smiles, particularly as the Chinese canteen contractor presented us all with 10 cheroots per head.

The men fared just as well. The Christmas meal laid out in the gaol kitchen contained 68, 000 rissoles, each man getting 10, each one different, besides soup, rice and an excellent nasi goreng (fried rice). For myself, much to others' envy, I succeeded in getting two ducks eggs at 3 each which I had fried with whitebait and sweet potato chips for breakfast.

Christmas cards done by the men flowed in. It was a period of real goodwill to all of us. We ignored the Japs of course.

The highlight, however, must be given to the Christmas pantomime *Twinkle Toes*, a superb production far and away above many London ones I have seen.

New Year's Day 1945 the Japs themselves gave us some little extras which gave us a good send-off. After the New Year they dropped the root vegetables from 180 lbs a day for the mess to perhaps eight or nine. The green vegetables appeared on a scale of 40 lbs for 180 people per week. In the middle of January the soya bean ration of two oz a day ceased.

Latterly, January to August 1945, I had to give up smoking altogether as the slightest irritant brought back the symptoms of my deficiency disease of the mouth.'

As 1945 dawned, civilian areas of Singapore were being pasted by Allied bombers, increasing the ugly spectre of untold more civilian casualties before the war ended. Still, bombs weren't the only thing raining from the sky: the Brits dropped 'Banana Money' too: $750, 000 worth of occupation currency overprinted with 'Banana Money will die with the Japanese'.

The population of Singapore felt freer now to venture out to their much-loved coffee shops and tea houses; nightclubs even boomed once more with the renewed optimism. This confidence spilt over into property investment, the market for shophouses over-heating. The Japanese tried desperately, and vainly, to rake in cash with a new lottery and a patriotic saving scheme. Whatever trans-nationalism had flared in the early years of the war was now clearly waning.

The Japanese collapse in northern Burma became a rout. Aung San started agitating the Japanese publicly in his frequent speeches, working behind the scenes to foster communication with communists and Force 136, the British-led underground resistance operating in Malaya.

On Borneo, the Japanese began moving the POWs to the interior on the first of three Sandakan-to-Ranau 'Deathmarches' on 28 January. A couple of weeks later, on 1 February, US rangers and Filipino Guerrillas rescued 513 American survivors of the Bataan Death marches. Within a week, MacArthur reported the fall of Manila and the liberation of a further 5000 prisoners.

On 2 February, American B29s sank Singapore's floating dock in the biggest attack yet on the naval base. Food, water, gas, electricity, equipment and drugs were all at their lowest ebb, contributing to widespread sickness and death in Singapore.

The rice yield for 1945–46 — now a mere 227, 000 tons for the region – was down by one-third of that of 1939–40.

On 1 February the rice ration was cut from 500 grams to 350 grams per day. By the middle of February it had fallen to 300 grams.

'Fortunately, however, palm oil as a lubricant in lorry engines proved a failure and, once more, from February onwards we had adequate supplies. I, having given up smoking, spent all my money on palm oil.

At the end of February a Red Cross ship came in and was unloaded by our own men. But we weren't to see anything from it for a couple of months.

My weight in February: 12 stone 12 pounds.'

March proved to be a bloody month, with 300 B29s raining fire on Tokyo, killing 100, 000. Japan closed all schools and children aged over six were ordered to war service. Mandalay having fallen to Slim's 14 Army, Rangoon was heavily bombed by the Allies. The Burmese Defence Army, previously loyal to the Japanese, did a U-turn and attacked Japanese in Rangoon, killing over 8000 before the war there ended. A decisive day in their history.

'My weight by now was down to 12 stone 8 pounds. Raids culminated with a huge raid on the docks with incendiary bombs by 200 fortresses on 29 March. Next day a plane came over and dropped leaflets saying that the invasion of Malaya was imminent. Then nothing except a few recce planes every now and again.'

Doubtless the Allied resources were focused on the Burma theatre, where the collapse of Japanese resistance through upper and central Burma, plus the Allies' heavy use of transport aircraft, allowed the Brits to push into lower Burma. Burma was reconquered by an overland offensive from north-east India — an operation considered unthinkable two years earlier.

On April Fool's Day it was no laughing matter when US troops landed on Okinawa, the last step before Japan mainland. On 7 April the *Yamato* was sunk off Okinawa. The largest, heaviest battleship ever constructed, she carried the heaviest armament ever fitted to a battleship: nine 18.1 inch guns (three inches bigger than the guns defending Changi). She was ordered to sail into this losing battle because it was considered dishonourable for the ship named for the ancient name of the country to surrender. Proof of this suicide mission is the fact that she carried only enough fuel for a one-way trip: 2475 of her crew were lost.

'My weight in April: 11 stone eight pounds and still dropping in spite of buying my share of whatever was obtainable, which cost me $250 a week.

It wasn't till Easter Sunday that we eventually received one parcel between 20 men from the Red Cross ship "as a special favour from the Japanese Emperor". Can you beat their cheek? Starvation was really with us; every grain of rice was so important that we all scraped our mess tins clean. One officer, however, scraped so much that he produced a razor-like edge to his spoon with the result that he cut his mouth; blood poured, but it was not for several minutes that we realised the real cause and burst out laughing.

At the end of April they made an issue of one parcel per head. These we had issued at one per 20 on Sundays and Wednesdays, when the four oz extra food cheered us up immensely, particularly the tastiness of salmon, cheese and sardines with the chocolate whipped into cream with the butter.'

On 29 April a Japanese kamikaze dived into the hospital ship USS *Comfort* off Okinawa, killing 28 including six nurses.

By the end of the month, Rangoon was evacuated by the Japanese commanders, who fell back to reinforce Malaya. The Japanese used the Thai–Burma Railway to withdraw (or 'advance to the rear' in their parlance). Within days, the British went ashore in Burma, south of Rangoon. General Slim re-entered Rangoon overland, but could not save British POWs who were summarily executed as the Japanese fled. The retreating Japanese also blew up St Philomena's convent, which housed 400 wounded and recovering Japanese soldiers, to save them falling into enemy hands.

'After April 1945, there was a reorganisation so I took the opportunity of moving out on to the balcony. Here I lived in luxury on my own, with the cool night air whistling round me and here the noise and smoke from the cookhouse didn't disturb me. Life was much more enjoyable.

By 1 May our rice ration was down to 250 grams for inside

workers, 290 for heavy duty and 150 for hospital. Pooling was forbidden by the Japs although it was done in the officers' area giving us a mean ration of 246 grams a day. Hospital patients got into such a state of debilitation that we insisted, Jap or no Jap order, in supplying them with part of ours. By May the fish ration was 4–8 grams per day, if sent in.

There was a cantankerous officer who was too proud to say that he would like a snail patty and then, when they appeared, demanded his share so we all had to give him a bit of ours. The next day a very special snail patty was prepared for him with a snail's head complete in the centre and the perimeter decorated with snails' horns, and — as a snail fries hard — these all stuck up magnificently. He couldn't face it and quickly retired.'

On 7 May, Germany surrendered.

'After the fall of Germany excitement grew intense, mingled with our disgust at the return at home to horse racing, Test matches, Henley and such other things. It made us feel very forgotten and I shall never believe that they were justified while 100, 000 British and 50, 000 Indian soldiers were starving in prisoner of war camps, and the native population of 1, 000, 000 were being starved and subjected to the most brutal treatment, while they still looked to the British Empire for deliverance.'

On 8 May Truman warned that 'victory is but half won'. The campaign against the Japanese homeland was stepped up, resulting in the 26 May devastation by B29s of Tokyo, including several royal palaces.

In Borneo, on 29 May the Japanese sent POWs off on a second Sandakan–Ranau Death march to the interior.

'Starvation by May had definitely set in. A fresh killed snake would fetch $20. Dogs were sold for $120. A scraggy hen for $100 or a runner duck for $80. A rabbit $150, snails were 50 cents each. Frogs legs $2 the pair. Not that I could ever bring myself to face such food but I quote the prices for interest. Thieving of food became common. To meet the cash demand which $30 a month couldn't possibly cover, every type of article was smuggled into Singapore under the eyes of the Japs and sold by dock working parties to the Chinese.

A pair of shorts would've fetched $8 if in good condition in

September 1944. In May 1945 the price had risen to $80. In May I sold my second pair of shorts and lived on the proceeds for a month ($45 only as they were badly torn). A Parker fountain pen rose from $40 to $400. Wristwatches of very mediocre quality by May were fetching between $1, 000 and $4, 000 while gold had risen to $200 a gram for 22 carat, scaled down to $115 for 9 carat. Those who sold their belongings cashed cheques for others at rates varying according to the world news of between 2/50 – 8 to the pound. (In Sime Road Internment camp the rate never rose above 3.)

It should be remembered that the black market, although openly forbidden by the Japs, was carried on by them secretly. To sell or buy a Jap guard's cooperation was essential. An idea of the rise in prices can be obtained when it is realised that the officers pay available to purchase in the canteen was about $25, 000; the sum actually spent was in the region of $400, 000. The men were by no means behind hand and run us pretty close.

To combat our hunger we now constructed little electric stoves in which we cooked everything we could lay our hands on, such as a piece of tap root, or sago flour. Perhaps a little *blacan* fruit with rice saved from our meals.

By the end of May all the Red Cross Parcels were finished and we saw no more for four months.'

From March to June, every POW in Changi Gaol had dropped at least a further stone in weight. One 6 foot 2 inch Australian reportedly weighed a mere 50 pounds, a Body Mass Index of less than 8.4.

On 4 June B29s fire-bombed Kobe. Ten days later the British liberated Burma.

'Till June I spent all my money on palm oil and, when possible, coffee beans. In June, desperately hungry, I sold my gold ring, the profits from which just lasted me out till food poured in after the Jap capitulation. If I hadn't done so I am certain I would have been desperate with hunger.'

Japanese shipping had been strangled. By July 1945 shipping was totally unviable outside of Japanese home waters, and approximately 40 percent of all home-water movements were sunk. Still only two-thirds of Japanese civilians considered the war lost.

In Singapore, Allied planes buzzed overhead nearly

everyday.

On 5 July, MacArthur reported the liberation of the Philippines and, within a week, the US Fleet began its first heavy bombardment of the main islands of Japan.

On 16 July amid high secrecy, the first nuclear weapon was tested in the desert south of Los Alamos in New Mexico. On 24 July at the Potsdam Conference, Mountbatten was told, in the most classified manner, of the plan to use something called an atom bomb. Churchill predicted 15 August as the surrender date for Japan. Meanwhile US carrier planes bombed Yokusuka naval base, destroying the last of Japan's navy, whose war-monger leaders were arguably the root cause of Japan going to war in the first place.

On 29 July Japan turned down an Allied surrender ultimatum: they chose 'prompt and utter destruction'. American mines sealed off all main harbours in Japan on 3 August, completely isolating the country. On 6 August, Hiroshima was bombed. The news ricocheted around the POW camps ...

'With the dropping of the first atom bomb we knew that the end was very near.'

On 8 August a meeting of the Japanese Supreme War Council was called to discuss accepting the terms of the Potsdam Declaration of UN, that is, immediate and unconditional surrender. The mooted attack on Nagasaki was only hours away. But the meeting was cancelled, incredibly, because one of the Council's members had 'more important business' elsewhere. The next day, Nagasaki was ripped apart by the equivalent of 20, 000 tons of TNT. Truman warned Japan: 'Quit or be Destroyed'. On 10 August, Japan offered to surrender, with MacArthur named to accept Japan's capitulation.

'Back in June a coconut cost $2 and we were lucky if we got one a week. By August 1945 a nut cost $6 and we would get, all being well, one between four, for a month. Cheroots rose rapidly from 5 cents each to $1/70 each. *Blacan* from 20 cents a pound to $25. Sugar from $3 a pound to $80 and was almost unprocurable at that price. Towgay beans went from 10 cents a lb to $24 per lb. Whitebait from 30 cents a lb to $60 a lb. All other commodities rose on the same scale.

Not only did the price go beyond our reach, but they were

almost unprocurable. Only our gardens and the black market saved us, and the magnificent ingenuity of our cookhouse officers.

Medical opinion is that within six weeks the death rate would have been appalling; it had already begun to rise alarmingly. Truly the atom bomb only just came in time.'

A caricature of Pilk in Changi, done shortly before liberation in 1945 by renowned artist Ronald Searle, a fellow POW in Sime Road camp. (Photo courtesy Pilkington family)

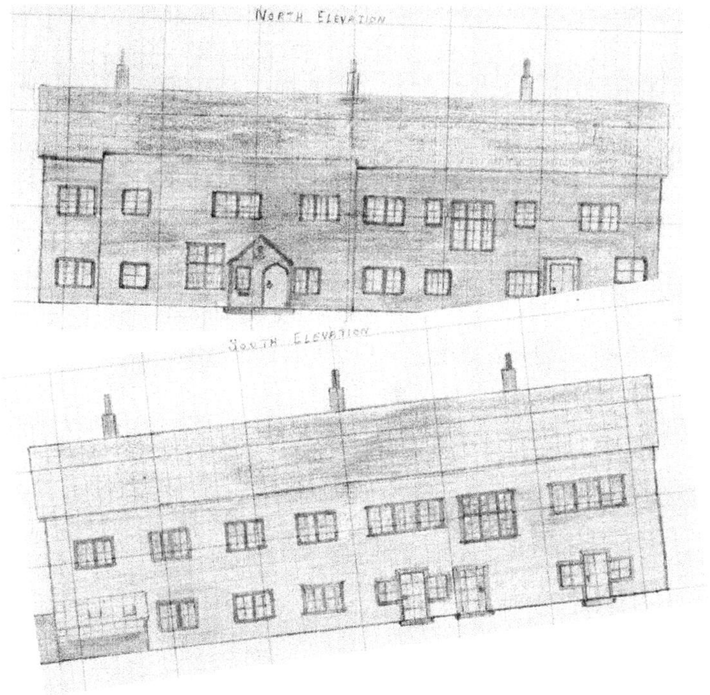

Dreaming of lavish homes and feasts helped the POWs pass time in Changi and other camps. (Courtesy of Pilkington family)

21 SAVED BY THE BOMB

Officers' Area, Changi Gaol, Singapore, 11 August–22 October 1945

'Since the dropping of the atomic bombs we had known that the end was very near and so, full of suspense, we waited hourly for the news bulletin that would announce the Japanese capitulation.

The Japanese "Gestapo" had for months done everything in their power to trace our wirelesses as they, quite rightly, said only two lots of people knew the real news in Singapore — the Japanese High Command, who wouldn't say, and the Prisoners of War, who denied all knowledge.

So rigorous had their inquisitiveness become that only three wireless sets were allowed for the whole camp by British order, to be operated singly on separate nights. These sets, a masterpiece of craftsmanship, were built by our Royal Signals out of bits and pieces and were so small that they were built into the wooden leg of *sharpoy* beds, and they were so finely tuned that the officer taking down the news had to listen with a stethoscope.

At about 2:30 am on the morning of 11 August I was asleep on my bed on the little balcony of our coolie quarter when an undue commotion from the Australian officers' hut over the road woke me up. Intrigued by the noise from the Australians, I walked across and looked in to see them feverishly sorting their belongings, and, when I asked what it was all about, I was told, 'We cannot tell you but you will hear soon'.

At 7:45 am when our Colonel Commanding woke up, I suggested to him that the Australians had some news. Colonel Harris went across and came back a few minutes later: "Gentlemen they have capitulated, but the Japanese themselves have not yet been informed".

No one spoke for several minutes, and then we all made a concerted rush for our little hoards of dried beans, fish, etc. and made fires to cook them on. By now we were really starving and food dominated our thoughts.

"At last" and "Thank God" were the only expressions heard.'

At home in Woolacombe, England, Phyllis would have been absolutely delighted and relieved by the telegram she received at this time: 'ALL WELL AND SAFE. LOVE.' Unfortunately, it was dated 30 January, 1942. A note was appended:

'In sending you the enclosed telegram the Postmaster General would like to explain it is one of a large number which have only recently arrived in this country from the Far East. As you will see it was handed in at Singapore over two years ago, shortly before the fall of the city and of course any statements made in the telegram may no longer be true. The Postmaster General can only regret that the vicissitudes of war have caused such a long delay and so make the message completely out of date. He feels that, despite the lapse of time, you would wish to be in possession this telegram. He realised that you have suffered, and may still be suffering, anxiety or sorrow as to the fate of the sender, and he asks you to accept his sympathy.'

'For several days the Japs would not admit that they had capitulated, working parties went on as usual, rations remained at starvation level and treatment was unaltered.

Over the wireless we heard that the Japanese Commander had said that Japan might capitulate but *he* wouldn't and that he would fight. Although we, used to Jap methods, realised that this was probably "face saving", we were still left in suspense.

Gradually, a change took place. Working parties were no longer called for, slight but steady improvement in rations, and change of behaviour on the part of our guards who became more lenient.'

Four hundred Super-Fortresses resumed the blistering attack on Japan on 13 August. But some within the Japanese Supreme War Council still wanted to fight on to redeem themselves on a home soil duel to the death. The Emperor himself was initially reluctant to sign because there was no guarantee that the institution of the office would be recognised and perpetuated. But eventually, thankfully, he agreed to 'bear the unbearable', in the process saving possibly another 1, 000, 000 Allied – and untold Japanese – lives. On 15 August the surrender was announced.

Then it was finally announced that the Japanese Commander, on orders from the Mikado himself, had capitulated. Even this bit of news was in our hands before our Jap guards were notified.

The war ended with the British having cleared Burma and poised for landings in Malaya, which was teetering on the brink of revolution. The Malayan Communist Party numbered around 5000 men by now, some of their units even boasting European liaison officers. But many did not want to show their hand until they knew which way things would pan out under Ibrahim Yaacob who, like Sukarno in Indonesia, was heading a nationalistic push for independence.

In northern states, including Pilk's beloved Negri Sembilan, Japanese commanders tried to broker deals with the Malayan Communist Party to make men and arms available to continue fighting.

As for Japan, the would-be regional masters themselves faced the rather ironic certainty of mass starvation in the winter of 1945–46, had the war continued. They were almost out of staples such as sugar and tobacco. Private citizens could get no petrol for their cars. The Japanese government put the A-bomb death toll at 190, 000 and rising still by 100 per day. It also appealed to the last patriotic *kamikazes* to give in.

'At once our OC Camp took immediate action. He asked for an interview with the Jap Commander in which he laid down his requirements. Firstly that Red Cross parcels — known to be in Singapore — should be delivered at once. Secondly that fish, meat, fish, vegetables and cooking fats should be delivered, as well as essential drugs for the hospital. Thirdly that we should be allowed a wireless. And lastly that we should now administer ourselves.

All the conditions were agreed to. Red Cross parcels poured in all day by the lorry load on 15 August. Some 28, 000 came in on one day, from the ship that had arrived in March. There must have been many thousands more as we know the Japs were selling them for their own profit.

Butter by the ton was produced for cooking. Fresh vegetables, meat and fish started arriving in driblets, but gradually increased to a river. From the Red Cross ship the Japs stole all the essential drugs including the insulin.

They tried to quibble over the wireless saying we already had one! Joint administration of the camp was agreed to, but they insisted on retaining their Jap guards.

On 21 August a Lancaster bomber at last appeared and flew backwards and forwards over the gaol dropping leaflets. These were most uninspiring, stating that help was on the way and that in the meantime we must eat up all our rice and make sure that we drank the water it was cooked in.

HINTS ON DIET DURING RECUPERATIVE LEAVE FOR LIBERATED PRISONERS OF WAR

As a result of the privations you have endured as a prisoner of war, you have probably lost weight, and it is natural to think that the more food you eat the sooner will you recover your lost weight and strength. But you must remember that your physique as well as your weight may be temporarily below par, and this includes your digestive system. Just as you need rest at first and your muscles require gradual retraining, so your digestive system requires rest at first and then retraining in the handling of the sort of foods you normally like to eat.

To get your digestive system back to normal as quickly as possible, a few simple rules that you should follow, especially if you are having trouble with your digestion, are given in the dietetic instructions below. You should show these notes and the following instructions to anyone who is giving you your meals, so that they can understand why you have to be careful about eating for a time, and what they should give you to eat.

(1) **Don't overload your stomach.** Avoid heavy meals, and instead, eat small amounts frequently. Try eating three light meals a day, with three snacks of the biscuits and milk variety —two between meals and one last thing at night.

(2) **Remember that your digestion is weak, and at first give your stomach foods easy to handle.**

Eat: Foods such as milk and milk puddings, eggs, cereals, toast or bread, biscuits, preserves, cake, and fish and tender meat if you can eat these without discomfort.

Avoid at first: Fatty or fried foods, bulky vegetables, raw salads or fruit, highly seasoned dishes, twice-cooked meats, pickles and spices, rich, heavy puddings and pastries, strong tea and coffee.

Beer and other alcoholic drinks are hard on a weak stomach, and you should take these very sparingly, if at all, for the first few days at least.

Canteen supplies improved immensely, both in quantity and quality, and with the money I still had left from the sale of my gold signet ring I could buy all I could eat.

On the night of 22 August we were informed over the wireless that essential supplies would be dropped to us by parachute and, next morning, bomber after bomber came over our Changi airstrip and rained canisters of supplies down.

At times ten to fifteen parachutes could be seen floating down at once. All these canisters were collected and brought into the camp and, when their contents (other than the medical supplies) were divided up, each man had two boiled sweets and half a cigarette per head!

Small though it was it was a token of the outside world and meant to us their weight in gold.

Two days later two parachute officers were dropped together with two NCOs of the Royal Signals with a transmitting set. The first officer to land, six feet in height and all of 200 lbs, when the Jap who met him tried to argue, picked him up and shook him like a rat. From then onwards the Japs fade out of the picture.

The sight of these armed men carrying tommy guns, sheath knives, revolvers, and hand grenades was marvellous. At once they took over command of the island and set up their wireless only to find that a crystal, essential to its working, had been left behind. We were able, however, to supply the missing part from our camp secret hoard.

These were the first well-fed Europeans we had seen for 3 1/2 years and to see them eat a whole Red Cross Parcel for a meal when all we could do was to eat one between four, was a sight in itself. Incidentally they condemned the Red Cross parcels as not being fresh.

At this time we were having one Red Cross parcel between four, to be increased to three and then two per day. When food was plentiful we soberly discussed the question as to whether leggee queues could be abolished and if the surplus food on our table such as butter and jam couldn't be kept whole to which we could help ourselves as our appetites dictated.

With the arrival of the parachute officers, the Jap guards were removed and replaced by our own military police. Bathing parties were organised and generally conditions improved. When we were released a party of British troops, some ten men, took over one of our trailers and dragged it up a hill; our 30 POWs could scarcely keep it moving.'

The Japanese troops in Singapore officially surrendered on 29 August. For some the shame of defeat was too much to bear – 300 officers held a sake party before blowing themselves up with grenades. A platoon of Japanese soldiers followed suit.

On 2 September the main Japanese surrender was signed in Tokyo Bay. Japanese banana money immediately became worthless, not exchangeable for Malayan money. Local 'Tiger' notes rose from 12:1 to 30:1 against the Japanese currency. The final issue of the *Syonan Sinbun*, Monday 3 September, ran the headline: **'World at Peace Again. Formal Surrender of Japan Concluded.'**

From 2 to 4 September reports of British shipping standing off Singapore came in hourly, and on 5 September the landing took place.

British warships led by HMS *Sussex* (later dubbed 'Hotel Sussex') appeared in the Singapore roads and the Japanese flags came down at the Cathay Building that day. At noon, Lady Thomas raised the British flag at Sime Road Camp. New World, Great World, Beauty World were ablaze in neon, with jazz bands and, most noticeably, English announcements for the first time in three and a half years.

That evening advance units of journalists and others found their way to the camp.

The British had lost 32 percent of their men in captivity. The Australians 33 percent. All told, including the Americans and the savvy Dutch (20 percent), 27 percent of POWs had perished under the Japanese. These figures generated banner headlines in the English newspapers.

The Times ran an article on 25 September headlined: **'British Rule Vindicated. Malaya's Great Welcome.'** The report from 'Our Special Correspondent in Kuala Lumpur' read: 'At every town and village on the road here from Singapore the people have built triumphal arches with welcoming inscriptions. The arches and the houses are decked with the flags of Great Britain,

China, and of the Malayan People's Anti-Japanese Army. Children still cheer wildly every passing vehicle, making the thumbs-up sign. There can be no doubting the genuineness of the welcome. The return of British rule has meant release from that ever present fear which hung over the people like infection in the atmosphere. Indeed, in Singapore one almost gets the feeling that the privations and indignities endured by prisoners and interned civilians actually raised the standing of the white race. Perhaps under the Japanese, the common bond of suffering and hatred for the invader tended to unite the communities. Although physically separated by barbed wire, European and Asiatic have probably been closer during the past three years than before.'

Pilk occupied this period of mental and physical limbo with more writing. This time it was *Paul's Own Story* – a collection of letters to the son he had not yet met.

'These letters, written as an infant's history of the war as it concerned him, occupied my spare time between the Japanese capitulation and our release by the British forces August 21 to September 10, 1945. They were never posted as was my intention as I was advised that, as censorship was still in progress, they might cause a great deal of trouble and might not even reach their destination.'

The manuscript amounted to 40 pages of lined paper scrawled on both sides with pencil. The first story ended, "With all my love and millions of kisses from your loving daddy." This was scratched out and replaced with, "Now good night and sleep well, your loving daddy". Clearly he was looking forward to playing that new paternal role once home. But when would that be?

On 21 September, resigning US War Secretary Stimson recommended to Truman that this horrendous war be designated as 'World War II. The label stuck. Meanwhile, to Pilk and tens of thousands of POWs still in Singapore, the end of the war seemed anticlimactic, a period of doldrums as they waited for something – anything! – to happen. Soon, the situation was explained to the remaining officers of 18 Division:

'Lord Louis Mountbatten said that full information regarding the Jap plans for POWs had been in his hands. All officers were to be assembled under the pretext of being moved to another camp and, on the way, we were to be machine-gunned. I later saw the machine-gun emplacements. All the units out on working parties were to be ordered down the tunnels they had dug when they were to be collapsed on them and petrol poured down and then set on fire to make sure.

To counteract their plans an airborne brigade was to be dropped on Changi Gaol, and another on Kranji Camp. These were to be assisted by a Chinese guerrilla division already organised. An armoured column was to force its way down from Port Dickson and Port Swettenham. Fighting was expected to be severe and they calculated that not more than 50 percent of the POWs would survive. Hence the delay in shipping to get us home as they had only made allowance for some 15, 000 POWs.'

Ex-internees and POWs numbered around 34, 000. With the tables now turned, there were half a million Japanese POWs in Singapore and Malaya.

'Ghurkha troops took over all the Japs as prisoners and, although Jap officers had been told they could keep their swords, this meant nothing to the Ghurkhas who decided that — as far as the Japs were concerned — it was their heads or their swords when they got their swords anyhow.

The Japs were allowed to pack up all the goods they could carry and were then marched to Johor, 16 miles away. When crossing the Causeway the Ghurkhas cut loose all their packs and packages and made them continue in shirt and shorts alone, dividing out the booty amongst themselves and the natives.

The Japs as a whole are a hard-working farming nation and no doubt in their own country are tolerably decent folk, but once abroad their ignorance, arrogance and cruelty can deceive no one. Their pitiful lack of organisation and intelligence is immediately apparent. They are still mediaeval, uneducated and bestial in their ways of justice and living.

I cannot say I loathe them; forgive them I never will or can, but at the same time I do not rank them in my mind with the German pre- calculated cruelties of Belsen. I don't think that even our Jap guards at their worst were capable of such crimes.

I think one should remember that the few educated Japanese never hoped, nor expected, to win the war. The others went blindly on the path and orders they were given; unthinking and uncaring.

I never wish to visit Japan or see a Jap again. My last view of them put to work under Ghurkha guards will be some palliative to our years of suffering.'

Perhaps the only Japanese thing of beauty to Pilk was his much- loved Japanese cherry tree outside his quarters, noted by Lady Louis Mountbatten on her visit to the beleaguered ex-POWs.

'Various lady visitors to the gaol before we left remarked that it was the only pretty sight in the camp, which pleased me as I was fond of that tree.

The shining light of Malaya of those years was the Chinese. They never faltered and from the first they had the Japanese beaten. They won every round of the Jap contest to dominate them. To us prisoners they were magnificent; they bought our goods to give us money; they kept our canteens going at prices far below those ruling in Singapore. Whatever they could do they did.

There was, of course, the seamy side of our prison life when senior officers abused their rank to get extra food; when officers had to be court martialled for petty theft; how officers — rather than face the Japs themselves — were at times found hiding, while their juniors got the beating up; the gang robberies by men from their officers. All these and many others are better forgotten.

In any case they are completely out-weighed by the hard and conscientious work, self-sacrifice and generosity of the greater majority of officers. No party of men can live cooped up for years without bickering and squabbling, so it is nice to think that on the whole we lived in friendship and real goodwill.

For myself I think myself lucky to have met so many first rate people, both officers and men, whose kindness and companionship made it possible to be comparatively happy over those grim years.'

On 10 October 1945, Phyllis received the news she was desperately waiting for — a Cable and Wireless Telegram 'Via Imperial', sent from Colombo, Ceylon. It simply read: 'SAFE IN BRITISH HANDS. HOPE TO BE HOME SOON.'

Jubilant scenes as POWs are released from Changi in September 1945 (NAS-SLV). Pilkington returned to the UK aboard the HMT *Sobieski*.

22 HOME AT LAST

England, November 1945–July 1982

And so it was around mid-November 1945 that 42-year-old Captain Hugh Pilkington made it back home to Phyllis and met his son Paul for the first time. Paul tells of that moment at Woolacombe, Devon:

'I had seen photographs of him I guess, but I remember the first time I saw my father. I'd just nearly turned four. My mother had obviously done a lot of preparation for it because he was coming home. All the photos were out, everything dusted off, and she explained who was who. My mother and I were very close and I knew everything was going to change, but I didn't know how. He turned up at the gate in a taxi. We knew exactly what time he was coming. I was wondering what all the fuss was about quite honestly. A father? Who needs one of those? So that was that. Bit of a strange affair; a bit strange for both of us.'

Soon Pilkington received a typed form letter from Buckingham Palace, signed with a flourish by King George R.I. on royal letterhead: 'The Queen and I bid you a very warm welcome home. Through all the great trials and sufferings you have undergone at the hands of the Japanese, you and your comrades have been constantly in our thoughts. We know from the accounts we have already received how heavy those sufferings have been. We also know that these have been endured by you with the greatest courage. We mourn with you the death of so many of your gallant comrades. With all our hearts, we hope that your return from captivity will bring you and your families a full measure of happiness, which you may long enjoy together.'

The mail also yielded another surprise: 'In October 1941 a friend sent me a copy of Mason's *The Life of Francis Drake*. This was forwarded by the Army to Singapore. I finally received it four years later when returned by the Army Post Office who were unable to deliver it as Singapore fell before the mail reached there.'

Pilkington was awarded three medals: the 1939–1945 Star, the Pacific Star and the War Medal 1939–1945. He also received the King's badge 'for members of the armed forces, merchant navy, Home Guard and Civil Defence Organisations who were disabled as a result of war service'. But there was still unfinished business to attend to ...

A typed 'Claim for Loss of Kit' template dated 19 April 1946 arrived from the Officer in Charge, Army Pay Office, authorising compensation for loss of civilian effects, sleeping bag, instruments and binoculars. 'A draft for £70 4s. 2d. is accordingly being forwarded to your bankers.'

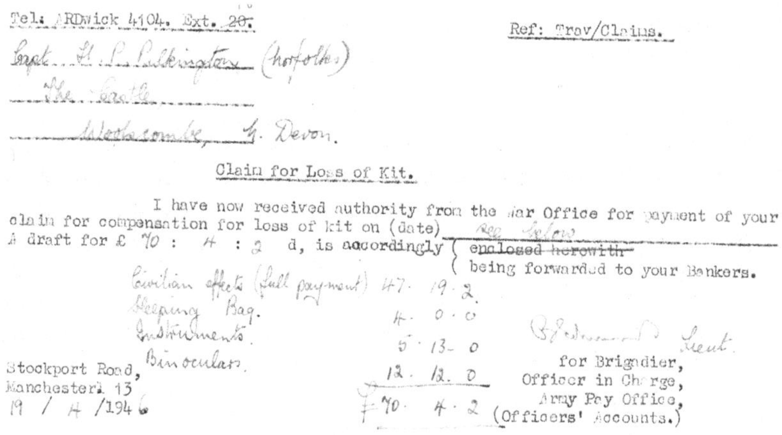

Tel: ARDwick 4104. Ext. ~~20.~~ Ref: Trav/Claims.

Capt. [illegible] Pilkington (Norfolks)
The Castle
Woolacombe, N. Devon.

Claim for Loss of Kit.

I have now received authority from the War Office for payment of your claim for compensation for loss of kit on (date) see below
A draft for £ 70 : 4 : 2 d, is accordingly (~~enclosed herewith~~
(being forwarded to your Bankers.

Civilian effects (full payment)	47. 19. 2.
Sleeping Bag.	4. 0. 0
Instruments.	5. 13. 0
Binoculars.	12. 12. 0
	£70. 4. 2

Stockport Road,
Manchester 13
19 / 4 /1946

[illegible] Lieut.
for Brigadier,
Officer in Charge,
Army Pay Office,
(Officers' Accounts.)

A battery of medical tests was followed by a letter of 2 June 1946 from DDMS, Southern Command: 'The disability has been confirmed as attributable to Military Service and the case falls to be dealt with under the Long Term Treatment rule'. Studies quoted by Brian E. Engdahl show that POWs who had experienced weight loss of more than 29 percent were at higher risk of developing depressive disorders, PTSD, and to a lesser extent anxiety disorders. A strong correlation between captivity trauma and psychiatric disability is evident. Captain Pilkington lost 30 percent of his 16.5 stone in captivity.

Then on 22 August a letter from War office released him officially from duty: 'Sir, Now the time has come for your release from active military duty, I am commanded by the Army Council to express to you their thanks for the valuable services which you have rendered in the service of your country at a time of grave national emergency, and to express regret that, owing to disability, it is no longer possible to utilise them in an active capacity. I am, sir, your obedient servant.' Given Pilk's disabilities, one wonders whether he was ever in fact 'released' or whether he remained in the captivity of his injuries. I asked Paul about his father's senior's appearance: 'He used to be propped up,' he told me, 'his shoulder in a sling. There was a deep impression in his shoulder; he couldn't move it very much. He was always pretty immobile. I don't think he played any sport at all after the war.'

That in itself would have been a crippling blow to Pilkington, the former rowing champ (he rowed at Henley versus Eton several times) and an all-round sportsman who represented his adopted state of Negri Sembilan in rugby, soccer and hockey. 'He did a lot of walking,' Paul continued. 'He used to take his dogs — he had a series of dogs, they all had Malay names — Tunku, Tuan, one or two others — and he used to speak in Malay to them. So presumably they understood Malay.'

Were there any thoughts of the family returning to Malaya post- war? 'No,' says Paul. 'My mum loved it out there, but Dad had had enough after fifteen years.'

Pilk attended the Norfolk Regimental Association Dinner in 1946, the invitation in *The Britannia* newsletter reading: 'The food situation is none too rosy just now, and our high hopes of a splendid feast are likely to be reduced to a crust and a nibble of cheese, but at any rate we can be reasonably sure of plenty of good fellowship, and a bar'.

A letter from Ministry of Pensions dated 30 November 1946, a full year after his return, informed Pilk, somewhat redundantly one feels, that: 'I have to inform you that the Minister of Pensions has decided that the disabilities 1/ effects of beri-beri 2/ gunshot wound left shoulder 3/ amoebic hepatitis 4/ perforated drum right ear on which your claim is based are attributable to war service. The combined degree of your disablement is based on an interim assessment of eighty percent and you are accordingly eligible for an award as shown below: total rate per annum £192 ... The retired pay is at the rate appropriate to the rank of Captain.'

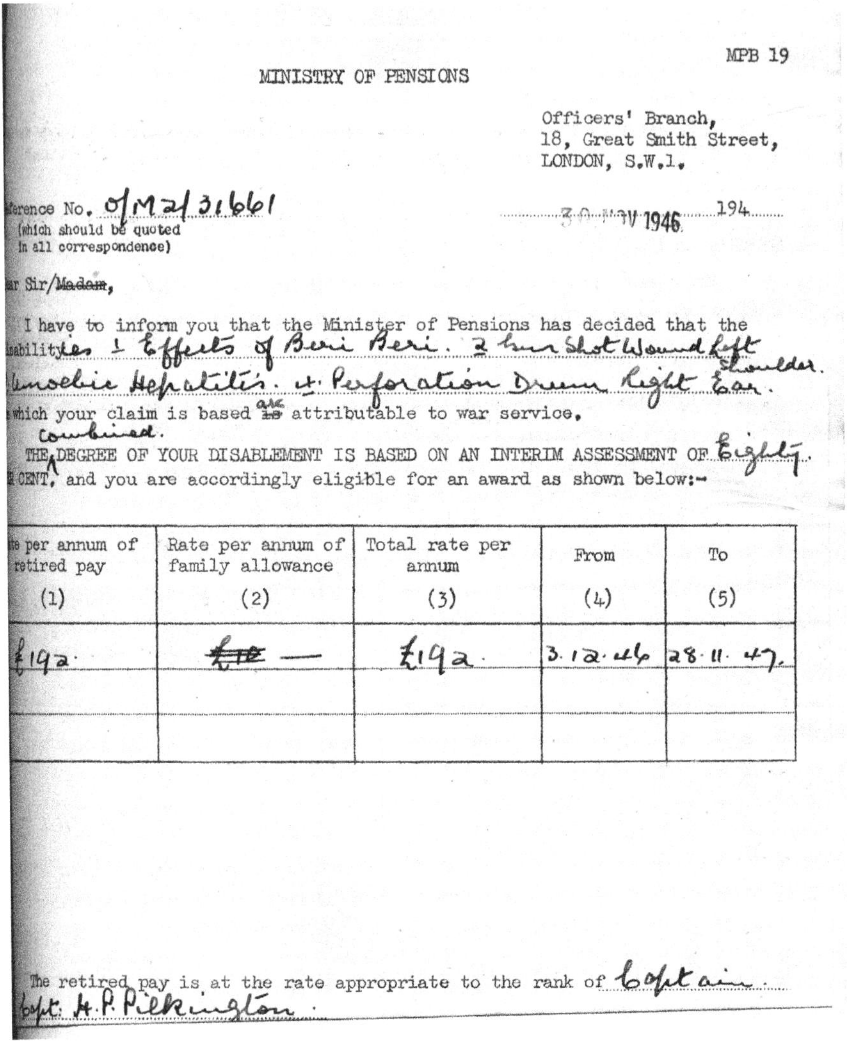

MPB 19

MINISTRY OF PENSIONS

Officers' Branch,
18, Great Smith Street,
LONDON, S.W.1.

[Re]ference No. O/M2/31661
(which should be quoted
in all correspondence)

30 MAY 1946 194

[De]ar Sir/~~Madam~~,

I have to inform you that the Minister of Pensions has decided that the [d]isabilit~~y~~ies 1 Effects of Beri Beri. 2 Gun Shot Wound Left Shoulder. [3 A]moebic Hepatitis. 4 Perforation Drum Right Ear. [on] which your claim is based are ~~is~~ attributable to war service.

THE combined DEGREE OF YOUR DISABLEMENT IS BASED ON AN INTERIM ASSESSMENT OF Eighty [PER] CENT, and you are accordingly eligible for an award as shown below:-

[Ra]te per annum of retired pay (1)	Rate per annum of family allowance (2)	Total rate per annum (3)	From (4)	To (5)
£192.	~~£10~~ —	£192.	3.12.46	28.11.47.

The retired pay is at the rate appropriate to the rank of Captain.

Capt. H. P. Pilkington.

Pilk had recurrences of malaria throughout his life, and his shoulder continued to plague him. He was hospitalised again for treatment in Roehampton in June 1947, during which time there was a royal visit by Lady Mountbatten. Pilkington recalled:

'Lady Louis recognised me as the officer with the cherry tree, which had been my pride. When she saw me as a POW I weighed 11 stone 8 pounds and was in rags. When she recognised me in

England I was 14 stone 10 pounds and well dressed and I was assured that she had not been informed about me beforehand.'

So did Pilkington ever regain the five stone — nearly one-third of his pre-war body weight — he lost during the war? 'Not far off it,' says Paul. 'He was always a big man — sixteen stone, seventeen stone. Dad was always obsessed with food and recipes, as evidenced by the hundreds compiled in his scrapbook from Changi.' Pilkington apparently cooked up a mean curry, always served with rice. (His culinary legacy continues with Paul also being a noted cooker and consumer of curries.)

Pilkington set about his post-war life with Phyllis and now two children, including Paul's newly-arrived brother, Rupert. 'He set up this business renting out agricultural machinery. He was contracted to mow air force bases, etc. Then he got the idea of drying out the clippings in a furnace and selling that as horse feed.'

But the war was obviously never far from his mind as in 1949 — having re-read his own diaries and letters — memories of further episodes and anecdotes were triggered, all of which Pilk had typed up and bound as *The Missing Years*. 'This record of the years 1941-1945 will, I hope, be of interest to you and the children.'

He studiously kept scrapbooks of architecture, transport — especially ships and trains — poetry, the history of his local Ilfracombe and Barnstable area, and birds. Obituaries and news features of friends and ex-POWs also figured prominently.

At the conclusion of the Siam Burma War Crimes tribunals, 111 of the 120 indicted were found guilty, with 32 capital sentences handed down. Approximately one Japanese was hanged for every 350 Allied and *romushu* death on the railway. Because many of these, such as Musso and Oswald, were only known by their nickname, it is difficult to determine who of Pilk's tormentors suffered what fate.

In 1952 Pilk received £30 10s. from distribution of Japanese assets: 'This amount also includes your share of the money allocated to the United Kingdom from the sale of the Burma–Siam Railway'. This effectively meant he earned much less than £5 per month for his work on the Railway. In contrast to Pilk's measly £30 share, in 1989 the US government paid US$20, 000 as compensation to each of its Japanese-origin citizens who was interned, in considerably better conditions, during the war.

Throughout the late 1950s, Pilk attended wherever possible the 18 Division Officers' Reunion Dinners held each October at the House of Commons. Multi-course menus invariably featured posh-sounding French dishes. Around this time, the family left 'The Castle' at Woolacombe. 'We then moved somewhere different each three or four years,' remembers Paul. 'At school, the talk was of how everybody's father had been a Spitfire pilot, or on a warship.'

Paul tells me of how he saw *The Bridge on the River Kwai* for the first time with his father at a special screening in London for ex-POWs in 1958. 'I think he got free tickets or something. On the way out he turned to the doorman and said, "What a lot of tripe".'

Ironically, December of that year saw the very last of the Japanese war criminals freed from prison, after liberal granting of clemency and commutations. No Japanese served a sentence longer than thirteen years. In 1972 Emperor Hirohito toured England (he'd previously visited 50 years earlier), where he met vociferous opposition. But that didn't stop Queen Elizabeth reinstating the Most Noble Order of the Garter decoration they'd stripped from him in 1941.

Initially, returning POWs might have masked their horrendous experiences by throwing themselves into work, family, friends. But as their lifestyles slowed down and they lost comrades or spouses, they had no option but to confront their terrifying memories. Prompted on the subject of life with Pilkington (who 'enjoyed a drink'), Paul refers to the disposition of his father: In those days the attitude was, "Bad luck, you survived the war, now get on with your life". I thought everybody's dad went to Changi. I thought it was normal. Not a day went by that he didn't mention the camps. The Japanese this, or Changi that. But he would never really *talk* about it ... never said anything, right till the end.'

That end came in July 1982 at the age of 77. It has been estimated that, clinically, the privations of POW camps and the Railway aged the surviving internees by ten to fifteen years (many in their twenties already had grey hair upon liberation).

As the English scholar and explorer Robert Burton — whom Paul got to know in Antarctica when he spent two and a half years there with the British Antarctic Survey — once wrote: 'If there is hell on earth it is to be found in a melancholy man's heart.' Clearly, Hugh Patrick Pilkington endured a living hell for decades after the surrender of Singapore.

Which puts me in mind of George Santayana who, in *Soliloquies in England*, wrote: 'Yet the poor fellows think they are safe! They think that the war is over! Only the dead have seen the end of war.'

Post-war, Pilk (right) gained back the weight he'd lost in POW camps, thanks to a lifelong enjoyment of spicy curries. He passed away in July 1982 at the age of 77.
(Courtesy of Pilkington family)

Epilogue

SIXTY YEARS ON
September/October 2005

Sixty years on from the end of the war, Asia presents an entirely different face to the world, in many ways attributable to the victories in battles achieved by the Japanese. The ramifications have long outlived their surrender in 1945.

Essentially, Japan was the catalyst for latent nationalistic movements throughout Asia. The Japanese showed that white man's colonialism was not the only model under which Asia and Asians could operate. Japan briefly toppled the Western powers, proved them not invincible, and undermined their prestige in the eyes of the local populations. There was no real coming back from this position for the once- mighty Raj, and other Western powers, who swiftly tried to reimpose themselves in lost colonies post-war. But the seeds of nascent Asian nationalism — Asia for Asians, without the overlording Japanese — had been sown ...

Within just two decades of the war, all prior possessions of Britain had become self-determining independent states (with only the ironic exception of Hong Kong — the first outpost of Empire to capitulate during World War II — which reverted to China under the terms of its lease in 1997). Non-British holdings followed a similar fate. Not that it was a trouble-free process. Each newly independent state spilled a lot of local blood on its road to freedom and independence; witness Vietnam. Some — the Philippines and Burma especially — still do. And in the case of Singapore and Malaysia, the battleground was particularly muddy as the citizens forged independent directions. Malaysia officially supplanted the Federated Malay States, the Unfederated Malay States, plus the former British colonies of North Borneo and Sarawak, in 1963. Interestingly, Malaya was called 'Malaysia' — a term in fairly common usage, unofficially, since the nineteenth century — by the Japanese during occupation, as proven by rare occupation currency coins minted in 1942.

Malaysia was now free of the colonial powers that had first imposed themselves there in 1511, when the Portuguese sacked Malacca. Initially Malaysia included Singapore, too. However, Singapore's agitation and ambition saw it removed from the Federation, much against Lee Kuan Yew's wishes, in August 1965. Malaysia's outcast southerly neighbour has gone on to post meteoric socioeconomic achievements. There is a keen rivalry (verging on enmity) between the two countries, akin to what one might find between, say, England and Scotland, or Australia and New Zealand.

Thailand always remained an odd-man-out in Asia; the proud never-colonised one. After the Japanese surrender in 1945, an interim government — in power only seventeen days — decreed that the country, renamed Thailand in 1939, should be known as Siam again. In 1949 it reverted to Thailand. It has its own internal struggles, however, chiefly in the predominantly Muslim provinces bordering Malaysia (the rest of Thailand being largely Buddhist). Separatist insurrections continue, adding to the thousands of lives already lost.

It is against this backdrop then, that Paul Pilkington, then 63, and I, then 42 (the same age as Pilk was when released), decided in October 2005 to retrace his father's memoirs — to not only see where he'd lived, fought, and been imprisoned, but to try and gain some sense, however marginal, of what it might have been like to be a POW; to experience his father's tortured past through the eyes of our privileged present. To add an element of hardship, we would consume the 'railway rations' as outlined in Pilk's darkest days up-country in 1943:

Breakfast – one teaspoon of sugar water, 1 pint of rice; Lunch – one pint of rice, 1 square inch of fish; and Dinner – pint-and-a-half of rice, with sprinkle of green vegetable.

Paul's participation would add a unique emotional dimension to the experience. 'It's not enough to read about it — you have to visualise it, take in the ambience,' he says. 'I am also keen to gauge the reaction to the Japs now.' His voice retains a Queen's English tone, thanks to his education in Devon, only somewhat diluted by his 36 years in the Antipodes (he arrived in Sydney in 1969 with a suitcase and a £10 ticket, the receipt for which he still has). But an exercise planned like a military operation this clearly was not. We were going to wing it; like the POWs who had no idea of their future fate.

To add authenticity, my 'Rules of Engagement' would include no change of clothes, no toothbrush, no media consumption and no contacting home (after all Paul, was sixteen months old before his father received word of his birth). Oh, and a thin straw mat on which to lay our weary bones each night in the worst possible accommodation we could find en route.

As those over 75 kilograms fared the worst on the Death Railway, I considered it prudent to engage a nutritionist to ascertain exactly what we could expect to endure, mentally and physically. My first major concern was about alcohol and caffeine withdrawals. Jo Beer, of Revitalise in Perth (and now a senior renal dietician at Sir Charles Gairdener Hospital), took us on.

'The alcohol and coffee withdrawals will possibly be the least of your worries,' she told us. 'You may crave them for a day or so, and you may experience headaches, but that will go after day two or three. However, if you do get dehydrated, headaches are a symptom. Other effects can be hair loss, cramps, bloating, diarrhoea, eye pain, rapid heart beat, bruising, slow wound healing, teeth grinding, anxiety and muscle spasms. Not a good list!'

Pilkington's memoirs don't mention much about water specifically, other than one litre per day when they were marched up to the Railway. So this was the amount we set out to consume.

'My most serious concern is that one litre per day is nowhere near enough for what you will be doing. At least 3–4 litres are more accurate. In that climate and with no fruit and vegetable containing water in the diet you will begin to suffer within a couple of days. The body can cope with a little fluid loss but once it gets to more than 2% the body reacts and here are some of the symptoms: dry lips, pale skin, sunken eyes, headaches, dizziness, not passing urine, heat illness and possibly vomiting. You will feel extremely tired and will want to lie down, possibly become disorientated, and, if that leads to no water being consumed then – as I'm sure you are aware – this can cause death. Sorry to put it so bluntly.'

Dick Lee, former 11 Division HQ dispatch rider and POW with D Force, came to our rescue, informing us that water up-country was virtually unlimited because they just boiled up more from the river when needed.

We also wanted to empathise with the loss of power, control, dignity that Pilkington and other POWs endured — the 'utter gloom' and 'mental hell' of being in Changi Gaol. One night at Ayer Molek Prison in Johor Bahru, Malaysia, does it. This 122-year-old prison was used by the Japanese during the Occupation, and, as the maximum security facility was about to be handed back to the State Government, I had pleaded hard, and ultimately successfully, our case to spend one night there. Unfortunately ... 'You are no longer Lloyd ... until you are released you are Prisoner 2505,' says barrel-chested Indian DSP Letchumanan, chest splashed with a row of service ribbons, four pips glinting on his epaulettes. Paul is 2506. We are fingerprinted, weighed, given a medical.

Letchumanan thrusts a sheet of typed paper under my nose. It is in Malay. 'This says you cannot sue the Prison Department or Malaysian Government for any damage and you enter of your own free will.' Hmmm, shades of the sign-this-or-else episode that led to the Selarang Barracks Incident. My collective life experience tells me not to sign it. Do I really want to throw myself at the mercy of the Malaysian prison system? We've come this far, though ... oh well, in for a penitentiary in for a pound. There is no chance to sign it 'Ned Kelly' or 'Mickey Mouse' as many POWs had at Selarang — I am instructed to affix my thumbprint.

Personal effects — money, passport, mobile phone — are put in a wax-sealed plastic pouch; inventoried. 'Your watch,' he motions to Paul. 'Watches are not allowed in case you are planning something.' I think of poor Pilk secreting his watch and ring when the Japanese stormed through the Alex plundering the patients.

'We will separate you so you feel the full challenge of this jail.' This lessens the impact of what it would be like sleeping cheek-by-jowl with fellow internees as the POWs had, but adds a fear factor.

My cell, redolent of a Spartan scout hall, is airless, stifling. Empty bar a half-inch bedroll and a bucket. The smell of rancid rat droppings is overwhelming. The door has twelve bars and double-meshing, a latch and a filthy big brass padlock. 'Once we close the door you are locked in till morning.' *KLLLAAAANNNGGG!!!* The door is slammed emphatically; the finality of the slam, gutting.

From the door I call across 30 metres of concrete, three walls, and three gates to Paul. 'It's not exactly the Ritz,' he shouts back. Clangourous reverberation makes conversation impossible. Sounds are my cellmate: the bellow of frogs, the chirrup of crickets, pigeons cooing, discordant Islamic chanting from a nearby mosque. The mournful wail of a siren winds up, then dies a slow, lingering sonic death. I am reminded of the Captain's thankfulness when the air-raid siren at the Alex took a direct hit. I understand that now.

I pace the floor. Crane my neck to see the sky: a full moon. A perfect bombers' moon, Pilk would have thought. I swipe mosquitoes away as I lie on my bedroll on the concrete floor. I slap at my stomach, awoken from sweaty slumber by yet another stinging cockroach bite. Cockroaches pour in – waves of reinforcements as though from trenches in the Somme – through drainage holes along the wall. The splattered carcasses of about 40 of their next of kin litter the concrete floor where the wrath of my prison-issue plastic slipper abruptly ended their miserable lives until I had given up.

My life at this point is equally miserable – the longest night of my life by far. I try again to imagine the POWs who, by my calculation, spent 1305 such nights in captivity, in conditions far, far graver.

In the morning, commander TTP Musa signs our release forms. A group of local school kids are being given a tour of the gaol. They gasp and titter when they see these two defeated-looking *matsallehs* (foreigners). It is ignominious. The same demeaning feeling the European civilians and POWs would have felt being marched out to Changi.

We collect our bags, and are officially free. Utter relief. 'Well, what other fun things do you have planned for us?' asks Paul.

'You are no longer Lloyd ... until you are released you are Prisoner 2505,' says DSP Letchumanan.

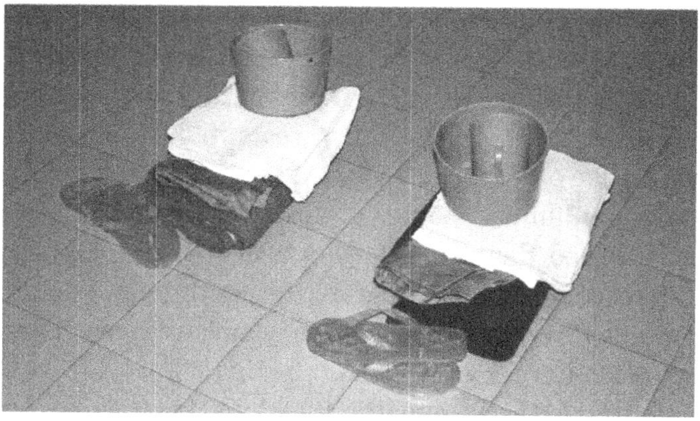

Our slippers, shorts, towels and toilet bucket at Ayer Molek jail. 'What other fun things do you have planned for us?' asks Paul, enjoying more rice.

We trudge across the Causeway, the same one so poorly disabled by the Allied sappers in January 1942, to Singapore. Of all the post- colonial success stories, Singapore is clearly a world-class example, and one I know well having spent a full decade living in the modern city-state.

It also the one place which treats its history as a national treasure. Unlike other former colonies — such as India which, upon independence, dumped statues of the former colonial masters in a river or banished them to a park outside Delhi — Singapore has embraced its colonial past as part of the continuum. So Raffles still stands proudly overlooking the river near Parliament House, and street names still reek of English bureaucrats over 40 years after independence. And, despite the wrecking ball taking out large swathes of the original city in the 1970s and '80s, amazingly much of the World War II infrastructure remains intact.

Post-war, Changi Gaol was renamed Changi Prison and resumed function as a civilian prison. It was redeveloped in 1980 into a more modern penal institution, then in 2004 the entire prison was demolished, replaced by a hyper-modern consolidated complex adjacent. The new high-tech maximum security prison holds 23, 000, usually making the international news when a foreigner is sent to the gallows for drug offences.

Today, only one section of the foreboding stone wall, 180 metres long, remains, its distinctive gate tower facing a huge Japanese school along Upper Changi Road North. 'Someone's clearly taking the piss,' I say to Paul. In a nod towards history and tourism, its main gate has been relocated from the side entrance to the front facing the main road. The re-created chapel and excellently curated museum have been relocated a kilometre or so down the road.

Singapore Armed Forces took over much of the Changi military infrastructure when the British pulled out in 1971. The former Roberts Barracks now stands within Changi Airbase, a high-security air force precinct built around the POW-built aerodrome. While the airbase is being upgraded, Block 151 (also known as Block W), the former dysentery ward and site of St Luke's chapel, is gazetted as a national monument so will remain as the home of the original 'Changi Murals' by POW patient Stanley Warren. Unlike

Pilk's thousand flies an hour, we spot not one. Inside St Luke's, ribbons placed by Japanese girls are conspicuously colourful and speak of remorse by the modern generation. The guest book had been signed by one David Evans, 13/9/05, Blackpool, England: 'LITTLE YELLOW BASTARDS!'

Not too far along the road is Selarang Barracks, headquarters of the Singapore Armed Forces Infantry since 1971. The squat buildings forming the quadrangle in which the Selarang Incident took place have been demolished, making way for a single larger administrative building and car park. The parade ground, about one-third its original size, is still used for 'square bashing'. But the original officers' mess, a glorious building in the English colonial style surrounded by swaying palms and lush lawns, still stands, housing a small World War II museum soon to be integrated into the base's main Singapore Infantry Museum. Our guide tells us stories of headless ghosts seen to stalk the mess.

Here, I have an illuminating discussion with one Captain Ang, a career soldier. Captain Ang's aunt was about sixteen when she was taken away by the Japanese, never to be seen again. How does that affect him personally in his relationship with Japanese?

'These are the bygones that we shall leave it as that, but the lesson is still there, lah. They taught us a very good lesson: the defence of the island, the country itself, does not depend on others. Every one of us has a responsibility. So that was a very wake-up lesson. But the Japanese youth don't think very much about all this World War II stuff. In their education it is brush off so I don't think they know very much about what is happening. As a friend, as a peer to peer, it's OK to have Japanese friends. Should not still be pointing fault on what is past, lah.'

I ask him whether perhaps Westerners dwell more on World War II than the Singaporeans.

'I don't think so,' he tells me. 'Because during Second World War everybody suffered, not just Westerners. In South-East Asia the Chinese as a whole suffered very badly, they just killed the Chinese as a whole. So, in a way, we still suffer but at this point of time we are willing to look one step forward. To progress rather than harp on the whole thing. Otherwise, both countries will never progress as far as what we have today. Japan plays a very vital pillar in the economy of South-East Asia.

'Generations like us we did not directly suffer under them — we didn't really live it so we are generally easy to forget them. I mean forgive not forget, lah. But the older generations are the ones who suffered no food, no water. Many of the relatives have been killed. That is something that is still very fresh in their minds. Many of them still live in that kind of nightmare.

'Recently we had a group of Australian and British veterans here. Sometimes it's difficult to conduct a briefing for them — you don't physically live in that type of environment as they did, so I end up just standing there listening to their stories. Some of them will be so emotional. When they are relating they cry.'

The detective work in trying to track down locations, such as the overgrown Tyersall Park within the grounds of the Sultan of Johor's property, was one of the most satisfying aspects of our mission. Next up we try to locate the former Sime Road camp. Sime Road is now one of Singapore's foremost residential addresses, fronting the exclusive Singapore Island Country Club where Japanese officers were known to brush up their game. Passing a lonely roadside pillbox, I ask a Eurasian-looking member in the car park if he knows where the camp was. He doesn't, but we get engaged in talk about the war. 'Any Japanese members here?' I ask.

'No, we've already chased them out once before!' he laughs. 'The Japanese are weird,' says the gent who's travelled there once before. 'But that's history. It was the generals convincing the Emperor to take over South-East Asia and side with Hitler. I don't understand that.'

Down the road I stop to ask an unassuming-looking Chinese gent in shorts and T-shirt, talking to some gardeners. 'Come.' I follow the middle-aged man across acres of lawn up to a sprawling old mansion.

'Japanese general's house before,' says the congenial Mr Lee, whom I later discover is a palm plantation magnate. 'Japanese dormitories,' he points to a raised area where sloped retaining walls and concrete footings are still evident.

The garden falls away to reveal more concrete footings and pathways. One ivy-clad concrete shed-like building still stands. 'That was the clinic,' Mr Lee says. The garden is a tropical wonderland of banana trees and ferns. No wonder Pilk described it as 'like waking up in paradise'.

Mr Lee is a charming gent who's clearly not worried by the snakes slithering across his lawn the way I am. He takes us around the monstrous house: 'The original owner was the chief surveyor, William. This was the hardest built house in Singapore – the concrete came from A.G. Dobb who was a concrete expert, and did all the works for Sentosa Fort and the Johor Battery.' He points out shrapnel scars still visible on steel railings. Having bought the house in the mid-'60s, he discovered there was a family connection: his brother-in-law Jeffrey Hudson was a civilian internee at this very camp. 'Now, he just laughs about it,' says Lee. He relates a funny story: 'The Japanese general used to play awful music on his piano. Somehow they got inside and dismantled the thing.'

An Indonesian maid, Mary, is industriously cleaning Mr. Lee's car, a Lexus. 'Toyota always made the best cars,' he says, making the point that he'd never bought a Japanese car on principle, until this one very recently. (Toyota is the most popular choice of car on the roads here, followed by Nissan and then Mercedes.)

Born in 1932, Lee has vivid memories of war-time Singapore. 'My father was beaten and bashed. We peeled the coat off his back but had to soak off his shirt, then rub sulphur in to treat it, fanning it to cool it.' His father's back remained scarred for life. A young Lee used to ask his father how he felt towards the Japanese: 'I try to forget: the guy who followed the order did a damn good job, the one who gave the order ... who knows where he is, so if I hold a grudge who loses? Only me.' Lee is an educated and worldly man – he and his wife Peggy are about to jet off to London for three months to catch up with dear friends – with clear opinions on the way things were. 'If Singapore was occupied by the Japanese Navy or Air Force, maybe not so bad because in peace time they travel to other countries and understand. But the Army ... animals. *Worse* than animals.'

So how does Lee feel towards Japanese today? 'I don't like them, I don't trust them, I have no Japanese friends.' What about doing business with them? 'Ah, that's different.'

Peggy has done her bit to perpetuate 'this vital chapter of

our lives' by maintaining the former camp grounds around her home. 'The young in Singapore have no perception of what we went through — the bombing, the onslaught coming so fast when the British were supposed to be invincible. Some of us were lucky to be evacuated to India, Australia, the UK.'

Attitudes to the 23, 000 Japanese living today in Singapore, plus the countless thousands of tourists each year, differ among the many I speak to. 'Maybe under Japanese we would have to bow three times to each other!' says Salleh, a paunchy, moustachioed Malay driver, laughing heartily at the thought of a war he is too young to remember. But taxi driver, Mr Kwan, has clearly lived through the horrors. 'FUCKING NIPPON! FUCKING NIPPON!' he shouts, totally enraged by the sight of a couple of young Japanese shopaholics walking down Orchard Road laden with designer-gear bags. In case they hadn't heard him, he waves an angry fist in their direction. I was taken aback by the ferocity of his reaction, knowing the usually guarded Chinese way. Perhaps he'd seen a baby sibling thrown in the air and speared on the end of a Japanese bayonet for sport. Who knows? Wars force people to see things they never should.

Intrigued, I speak with Misake, a Japanese stewardess flying for Singapore Airlines, who has been living on the island for nine years. 'Singapore have big Japanese population but I don't mix with them ... too busy,' she says in an almost-local accent. She's good-humouredly up with some Singlish slang terms I test her on to loosen her up for the more serious right jab: what does she know about the Second World War? 'Before, a war thing between Japanese and Chinese.' Any problems being a Japanese in Singapore now? 'No, very long time ago.' She's almost affronted by my insinuation. 'In Singapore it's very international, many nationalities living together. It's good, no problem.'

Among ex-POWs, feelings are still polarised. Dick Lee has travelled to Japan since the war, and even stayed with a former IJA guard he'd been put in touch with.

'You've gotta put things in their right perspective. Them bastards that we had to put up with out there, they're not the same Japanese as you're talking to now. The Japanese civilians were under the thumb of the military, Tojo and his bleedin' crowd. And the ordinary Japanese, you couldn't wish for nicer people. You can't hold nothing against them. The only time I feel bitter is when I talk about it, I see the faces of them bastards that were our guards, 'cause I had a few beltings from them and all.'

John Miller, ex-Australian 2/10 Field Regiment, has visited Japan several times, because his daughter lectures at a university there. 'I have absolutely no problems with the young Japanese,' he tells me.

'World War the Second,' Singaporean cabbie Phillip Wee shakes his head, not using the more colloquial 'Japan Time'. 'So many stories, never end. In those days we can only eat tapioca and sweet potato. Now very expensive!' Pressed, Mr. Wee confesses to still hating the Japanese —'Still don't like' — even boycotting their electrical goods and appliances. 'But new generation different ... they send their students to Singapore to communicate, send students to pray at Beach Road Cenotaph, to the dead. Together they pay respect. Young students have already forgotten. Previous generation has nothing to do with this generation. Japanese PM [Koizumi] very stubborn ... now they got problem with China and Korea.'

So did Singaporeans learn a lesson? 'Yes, watch the Japanese if they build up.'

Another who lived through those rigours is the war hero, Elizabeth Choy, who died aged 96 a year after I met with her. 'I lost seven inches off my waist — most ladies would be happy to do that!' the dignified lady jokes of her time incarcerated in a tiny cramped Kempeitai cell for 193 days for helping out British internees. (The Kempeitai were disarmed and disbanded in August 1945.) Photos in her youth depict her as the archetypal cheongsam-clad Shanghai Siren: a very fetching woman indeed. During internment she was subjected to electric shocks in front of her husband. 'Even till now, anything to do with electricity, like microwave oven, ooh ...' she shudders. Suddenly 60 years ago is as fresh as yesterday.

By contrast 'KC', a well-travelled twenty-something Chinese Singaporean, is reluctant to share her views as she is very enamoured of Japan and feels Elizabeth Choy's views may not be reflective of the general population.

'I have travelled to Europe twice,' KC tells me, 'and what

turns me off is the fact that some locals whom we come across treat us as second class citizens because of our skin colour and inability to speak their language — this I hardly experience in not just Japan, but many Asian countries as well. So I'm more likely to side with the Japanese than Europeans — I feel that I receive more respect from the modern Japanese than the biased Europeans.

'Modern Japan has attempted to soften most Asian nations' negative attitude towards them through economic, especially in Indonesia and Cambodia, and cultural factors, especially in Taiwan, Singapore, maybe Malaysia and Indonesia. Singapore teens follow Japanese fashion trend closely.'

But raw historical sensitivities linger on in East Asia. China and Japan seesaw in relative influence. Resentment has simmered for nigh on 100 years.

January 1998 saw Premier Ryutaro Hashimoto officially apologise for Japanese cruelty to Far East Prisoners of War by shaking hands with Tony Blair. April 2005 saw PM Koizumi offer the most 'deep remorse and heartfelt apology' to China and South-East Asian nations for its 'tremendous damage and suffering' caused. But, placatory as these words are, there seem to be residual misgivings. For this is the same man who vigorously defended his visits to the Yasukuni Shrine (where fourteen 'Class A' war criminals, including Tojo, are buried and the first train through Konkoita is displayed in all its polished splendour), despite the setback to Japan's national interests by so doing. (I visited the impressive Shrine in 2018, and a diorama on one wall helped me understand clearly why Japan lost the war: the sheer size of the war front was insuperable – there was no way their supply chains could have kept pace.)

At war's end, the occupying Americans dissolved the Imperial Japanese Army. Article Nine, enshrined in Japan's redrafted constitution in 1947, has it that the nation will forever renounce the use of force. But Japan still wants its 'Self-Defence Force' to be recognised as a military force. Japan's inscrutable intentions still evoke distrust.

Next stop, the Alexandra Hospital. Its orange-red terracotta tiles and cream walls peer from behind a lush screen of greenery. Bar the enclosing of the verandahs around 1971 (when it was handed over to the Singapore Ministry of Health) and a coat of paint, it presents physically identical to how it was in 1942. Mr. Ang Kee Seng, a clinician and longest-serving staff member, shows us around. In this lobby, poor Lieutenant Weston thought it was the smart thing to walk out with a white sheet of surrender aloft. On this very floor, his blood once seeped.

At the corner of the wing we find the stairwell on which those two shells burst on 15 February 1942. 'Original,' says Ang of the highly polished stone stairs. Any damage has long since repaired. 'Original lights, original hand rail,' he says of the thick, curvilinear wood. If only inanimate objects could speak. Closed in now, and with the casuarinas flourishing at full height outside — unlike the saplings they would have been during hostilities — it is impossible to triangulate the various sights Pilk referred to. Singapore city is to the east or south-east, but trees and high-rise Housing Development Board apartments (in which 90 percent of Singaporeans live) block the view in that direction. To the west, Ayer Rajah Expressway is now a massive multi-laned conduit with exits and flyovers, overshadowed by towering residential and commercial towers.

Bushido be buggered! How could Pilkington witness such dastardly deeds as occurred at the Alex yet be so dispassionate and forgiving? How could he feel the Japanese were blameless in these actions?

Back in 1941, Bill Frankland, RAMC, had flipped a coin with his best friend Captain Tom Smiley to see who would be assigned to the Alex. Smiley won, while Frankland remained as Officer in Command at Tanglin Military Hospital. 'Tom smoked, and the last present he was given before leaving was a silver cigarette case,' the spritely 93-year-old tells me [Actually false: it was given to Smiley by his fiancee before leaving England.] 'The bayonet struck it, that's why he escaped. He said: "You see, if I wasn't a smoker I wouldn't be alive!"' he laughs.

After being dive-bombed on his dispatch motorcycle Dick Lee was transferred to Roberts Barracks Hospital then — when that was bombed — to the Alex. He fills in some of the emotional blanks of the massacre: 'Our officer says, "If the Japanese come in, you won't understand them, they won't understand you. But everybody must stay where they are ... put your hands up, let them see that there's no resistance." Got the set-up have ya? Got the idea? You understand me? '

His voice rises and breaks somewhat at this point.

'Can you imagine the state of people in that bleedin' place? People with legs off, arms off, backs blown out. So many wounded, like a bleedin' slaughterhouse. That's the part that's bitter in me to think what they done in there.' (Dick didn't know it, but that utterance inspired the title of my next war book, and set in train a 14-year research deep-dive into the Alexandra Hospital Massacres, resulting in the publishing of *A Bleeding Slaughterhouse* to coincide with the 80th anniversary of the fall of Singapore in 2022.)

I tell Dick I assume the massacre was the lowest moment of his life.

'No, I would say the Railway was the worst experience I've 'ad. You're watching people up there dying of all these bleedin' diseases and the suffering, rotting away with these fuckin' ulcers in their legs, and the pain and suffering that they were going through, and you couldn't get an Aspirin, couldn't get a jab. That's suffering. Up in that railway you're talking about weeks and months and they're dying like flies with dysentery, malaria, ulcers, starvation, brutality. It's a different thing entirely. The hardest part was not knowing when it was going to end. When's it gonna finish up?'

This predicament was probably not helped by frequent reference by the Japanese to this being a 'hundred year war'. Dick sadly succumbed to TB, suspected to have been contracted in the camps but dormant all these years, soon after our final interview in 2008.

An Italianate memorial fountain gurgles adjacent the hospital's administration block. It is supremely serene with its trickle of water, chirping birds, and a spray of orange-bloomed creepers. Young lovers while away the evening on benches, suddenly self-conscious at our presence. Paul reads the plaque dedicated to the massacre's victims. The hospital will be moving to a larger site in 2009. It is my hope that at least this fountain and plaque will remain as a memorial tribute, but the hospital is making no such commitment.

Mutaguchi, Commander 18 Division, was arrested but never

tried for this massacre as a war crime. However, in a pre-trial assessment handed down by Colonel Kerin in 1946, it was said that no-one identified the soldiers that did the killing, and evidence against Mutaguchi being responsible was 'presumptive' as he was not present and it was not shown that he was cognisant of what took place. That Indian troops had fired from the hospital buildings was mitigating. Kerin concluded: 'I do not see that a case can be made against him.' Lieutenant General Nishimura Takuma, Imperial Guards' Commander, started his Malayan campaign with 300 killed at the Bakri massacres. This was shortly followed by 145 seriously wounded slaughtered at Parit Sulong, their bayonets hardly having time to dry before they reached the Alex. Nishimura was hanged for his role in Parit Sulong, thus possibly avoiding trial for the hospital episode, too.

Qingyan, a 30-something mainland Chinese anaesthetist practising in Singapore in 2005, shares her views on the atrocities by the Japanese: 'No, they were not beasts; that is not being fair to beasts. How can soldiers...' Tears well in her eyes. 'I hate the Japanese government, their country, because of their history. I have Japanese friends but they are different ... one to one they are OK. But I do not, will never, buy any Japanese products.'

Post-war, the victims' remains were reburied in Kranji cemetery where 23, 943 names are etched into the memorial wall. The cemetery is at the site of the former Kranji POW camp, overlooking the Causeway, where a small cemetery was started by the prisoners. Post-war, all recovered Allies who died in Malaya and Singapore were interred here, chosen over Changi as that site could not go undisturbed in the more developed area.

I chance upon several stones whose names jump out at me: Captain P.M. Westgate of the D Company, 6 Royal Norfolks, who had paid full price for bananas because he believed an officer shouldn't bargain. Pilk had last seen him in action when they were overrun on Bukit Pelandok. Dead aged 27. And Pilk's commander, malaria-ridden Lieutenant Colonel I.C.G. Lywood, who'd visited Pilk twice in the Alex, killed on '14–15 February'. Such was the confusion in the hospital no one could narrow it down to the exact date. Dead aged 42. My age then. Further along, the headstone of Lieutenant W.F.J Weston, who'd gone out of the Alex to raise a white sheet in peace. Dead aged 27. Corporal Bob Veitch, the anaesthetised patient on the Alex's operating table. Dead aged 28.

Then there's Private Rodney Breavington, 38, who showed such bravery when executed on Changi Beach. (It is assumed he was given a field promotion somewhere along the way as he was known widely as a corporal.)

At Kranji, I speak with Stan Sharpley, formerly of the 2 Loyal Lancashires, garrisoned in Singapore from 1938, who fought alongside Pilk's Norfolks around Yong Peng: 'The reason I'm here today is to see me old pal Titch Creston's grave. I was about 50 yards off when he got shot. Our CO had given us orders to withdraw because the Japs were coming over Yong Peng Heights. So I said to young Titch, "Get on the back of this bike". "Oh, I'll be alright," he said. I never saw him again.'

Another veteran Roy Miller, ex-Royal Navy, found the graves of three friends: 'It's lovely to be here ... wouldn't have missed it for the world,' he tells me. 'I feel as if I've communicated with them, experienced comradeship, found them. At least I know where they are.'

Dick Lee has come back to Kranji each of the four times he's returned to Singapore.

'I go back to pay respect to my comrades. And I just walk along the rows, don't recognise any particular names. I think of them days and pay my respects. Stop at a grave here and there, and salute the stone as a mark of respect. I keep telling myself I could've been there as well. You thank your lucky stars we are the lucky ones, amen. But I never forget. There's hardly a day or week goes by ... my thoughts always go back to the war, and Malaya and the hardships of being a POW. But you can't tell enough of it, you know. You can't get into the nitty gritty of it, because it's very difficult to tell a thing unless you've lived it. You've got to live the life that we have to try and understand it more.'

It's apparent that there is no escaping the tortuous mental incarceration that endures till now. After 60 years, many are still stuck on that island. Whether those who made it back were the lucky ones or not is debatable: the atrocities still rage in their heads. Long after the final shots rings out, long after surrender is signed, wars continue to exact a deadly and debilitating toll. Not just on returned servicemen, but with their spouses and children, too. And their spouses and children in turn (as I've noted leading military history tours through Asia in the past few years).

One thing I've never understood is why, upon the surrender of the Japanese, few POWs exacted any retribution for the ill-treatment they received. Dick Lee weighs in heavy on whether he would've liked to have done something to the guards. 'Yeah, I would love to have got stuck into that little bastard who give me a right doing with a bamboo stick because I had a bleedin' sore back for weeks. Thought he'd broken my back, little bastard. But then you quickly forget about all this wanting revenge, you were so pleased to be free again.'

Me? I think I would've ripped off the head and defecated down the neck of any guard who'd treated me as sub-human. But consider Pilkington's measured reaction, ranking the Japanese at their worst not capable of lowering themselves to the calculated cruelty of the Germans.

Statistics tend to belie this: over a quarter of all Allied soldiers taken prisoner by the Japanese were killed or died in captivity. One in three Brits, one in four Aussies, roughly. This contrasts with just four percent who met a similar fate having been captured by the Germans and Italians.

I speak about the fall of Singapore with Lian Hui, a slender short-haired officer with the Ministry of Education. 'The fall of Singapore is more hyped than the liberation, for us. That was always my personal thought but I didn't actually realise it until a work colleague mentioned it. Interesting huh?' Evelyn, a tanned PR officer from Sentosa Island (known as Blakang Mati during the war), offers her explanation as to why: 'Because Singapore had to fall in order for us to go on and achieve what we have achieved as a nation'. I am left with the distinct impression that Singapore's liberation date of 12 September is a more significant anniversary to Westerners.

Going home meant different things for different soldiers. In those days, there was no counselling, no diagnosis as post-traumatic stress disorder. Everyone was told not to talk about it and just get on with life. Worse for those from the Far East was that there were long delays in getting them back home and the European war was already long over. Many — listed 'missing presumed dead' — arrived home to find their girlfriends had married someone else, compounding their misery and detachment. Dick Lee was one such. In other more confusing cases, one POW's fiancée had two children with another man but was still waiting for him, and another POW's girlfriend had married his father!

One of those lucky to make it home from the Death Railway was Cyril Gilbert, a former sergeant with Australia's 8 Division: 'For two years it was wine, women and song, and I never sang a note!' he tells me, as he nurses another cold Tiger Beer, adding: 'I would always be looking in the fridge for more food, having meals in between meals in between meals. I survived this lot, now my next goal is to live to 100.' At 85, he is the only World War II veteran on the executive of the Returned Servicemen's League in Australia, and I sense he'll do it easily.

Teetotaller Bill Frankland — who had spent an afternoon smashing whisky and gin bottles at the Singapore Cricket Club ahead of the Japanese arrival — took a characteristically more sober approach:

'When we arrived in Penang we went to the RAF, and a nice Irish Colonel said 'I know what you want, I'll be back in twenty minutes.' And he bought a bottle of whisky. What he didn't realise was that we were all starving — we wanted food! He was the only person who took a drink from that whisky. When you are starving you don't want alcohol.'

A funny aside: I was in a discussion in 2019 with Tan DinXiang, the communications manager at the Alex. We were wanting to clarify some point and I said, 'There's one person who would know the answer to that – Bill Fankland, but he was 93 when I met him, and that was 13 years ago ...' We googled Frankland, and amazingly he was still practicing in Harley Street aged 106!

Dick Lee recalls leaving Singapore on a troopship:

'I looked at that place and, as beautiful as the Far East looks, I thought if I never ever see this area again I wouldn't miss it. When I came home a lot of people that I met at reunions just shut up, they didn't want to talk about it. I was always free and easy. But I was a bit down in the dumps when I first came home. I wasn't too bad with my brothers, but I found I couldn't mix too much at the beginning. I found it hard to mix with people, especially women, because they was difficult. I couldn't sleep on the bed, I slept on the floor. For a few weeks I couldn't get comfortable in a bed. So strange, everything.'

One of the burning questions I have is that, after an estimated 3800 servings of rice — often weevil or maggot infested — whether anyone would want to see another grain of the stuff again.

'Oh yes, I have rice a couple of times a week,' says Dick. 'Even on the ship coming home, the first sweet that came up was a lovely bucketful of beautiful rice pudding, with all the eggs in it and sugar. Some blokes went to the porthole and slung it out. "You're effing mad," I said, "this is not like the bleedin crap we were eating there."'

What does South-East Asia mean to them now, I wonder.

'Singapore's changed an awful lot to me,' says Stan Sharpley, 'but the people are so friendly wherever you go, into a supermarket, into a restaurant, go into a pub: "How do you do, sir, are you alright?" Great they are, smashing people.'

Bill Frankland says his feelings and emotions on coming back for the first time were, 'Strangely, none. Everything had changed so much.' He was far more worried going up to Thailand where he'd never been as he spent captivity on Blakang Mati, amazingly seeing the water only once in all his years there.

But if South-East Asia was so horridly ghastly during the war, why the lure to come back? Despite his love of Malaya, Pilk never returned. Dick Lee did.

'As years go by, you start getting curious and wonder,' he tells me. 'And I thought about that bleedin' railway job up there as well. And I got the inclination I wanted to go back. Thirty years had gone by. 1974. And I went back to Singapore and went up on the railway, to the areas where I worked. It was a strange feeling first time I come back on my own.'

He tells me a story of befriending two families (one Chinese and one Eurasian) in that 'funny atmosphere' once Singapore had been liberated. 'I thought of these people whose name was the same as mine, Lee. I was going to try and find them again: I found them after 30 years.'

Rear-Admiral Guy Griffiths (Retired), an Australian survivor from the HMS *Repulse*, reflects more philosophically when we talk at the dedication ceremony for the *Repulse* memorial at the former naval base site, now the massive Sembawang Shipyard.

'Every time I go to a veterans gathering I think how futile conflict is. You lose a lot of people. In the end, what did that all solve that couldn't have been solved with talking or diplomacy or different understanding and it goes on till today, right now. War is a silly pastime. It might appear to be a great adventure. I mean, we were thrilled to leave college as midshipmen and go overseas into the war. But conflict is a futile pastime.'

Amen. Hughie Clarke made it easier to visualise by claiming that a life was lost for every sleeper laid on the Railway, and Ronald Searle worked out that if the men who died building the Railway were laid end to end they would roughly cover the 273 miles of tracks built in that year. That 12, 399 Allied POWs and approximately 90–100, 000 *romushu* Asian labourers died in the Death Railway's construction is truly horrific. The numbers are still too huge to digest. That anyone survived, truly amazing. Looking at this another way yields yet another stark statistic: for each Japanese working on the Railway, there was one dead POW.

So have we learned anything from the Pacific War? Guy Griffiths weighs in: 'People who read history always talk about learning the lessons. But for God's sake apply them if something crops up in the future which is something not entirely dissimilar to what you've been through before. When the situation arises they forget about the lessons and deal with it as it is.'

After a major redevelopment at Changi Gaol, only the distinctive gate of the original building, built in 1936, is left.

Elizabeth Choy, OBE, aged 95.
'I lost seven inches off my waist — most ladies would be happy to do that!' she joked of her 193 days of solitary confinement at the hands of the Kempeitai. (She died in 2006.)

The beautiful art deco facade of the Singapore railway station at Tanjong Pagar, which had served Malaya since 1932. (It was sadly closed in 2011.)

Right: Paul and I boarded here -- in the same spot his father and thousands of others of POWs -- for the 1800km trip north to Ban Pong, Thailand.

Of the three countries where this story unfolded, Malaysia is clearly the most apathetic toward the war's history. One seemingly urbane youngster, Shahir — who professed a love for the *National Geographic* channel and watching documentaries — floored me with his attitude and awareness of World War II: 'Forget all this,' he said. 'Nothing important happened, lah.' A 20-something advertising executive, Shannen, is originally from Batu Pahat, a major centre for the Allies and Japanese during the war. I ask if there is or was ever much talk about the war years in BP. 'No, I have not really spoken to the older generation on this,' says the confident youngster. I tell her a few of the local wartime episodes. 'I am really surprised that BP was so scarred by World War II.' She also has no knowledge of the events at Parit Sulong, a mere twenty kilometres up the road.

A clean-cut chubby chap and keen golfer called Mukri gets onto the topic of colonial masters: 'Brits better than Japs, but now Malaysia better. Japs silly — they cut off heads. Brits build buildings,' is his pithy view.

And one of those buildings was the 1930 Public Works Departments barracks alongside the long hump-backed bridge at Parit Sulong, where Lieutenant Colonel Anderson's injured troops were holed up and slaughtered wholesale. The new bridge, completed around 1995, is a thick-set concrete structure similar to its famed predecessor. Vines and the jungle have reclaimed much around them, and the site is forlorn. Rotting rafters are collapsing. Its peeling walls are silent but seemingly wanting to speak. A couple of red poppies have recently been placed on the veranda. As we poke around the back, a voice calls out: 'Hello my friend!' It is Ariff, a short shaved-headed Malay who grew up and still lives in the kampong behind. Dressed only in jeans, he shows us around, pointing out the 'killing field' area which was burned and cleared two years ago. 'They are here — Indian, English, Australians ... definitely here,' he says, pointing to a pit in the dark soil in front of the vine-covered remains of the police lock-up. Australian military historian and author Lynette Silver is convinced the empty paddock behind holds all the never-recovered remnants of those massacred, and she is currently seeking approval to excavate the site. (She was also instrumental in having a commemorative memorial unveiled there in September 2007, a ceremony which I attended with relatives of the decimated 2/29 Battalion.)

Any ghosts? 'Ooh!' Ariff shivers, running a stubby finger down his goose-fleshed arm. 'When I am young ...' He doesn't elaborate but recalls seeing many, many bones. He points to a concrete well, indicating a beheading movement with his fingers.

Speeding back toward the visually prominent Bukit Pelandok defile, I feel the distance acutely. 'Gee,' I tell Paul, 'your dad covered a long distance on that first day. That's a rigorous journey just getting to and from the various battlefields. And it would feel doubly long if you were getting your arse shot off in retreat, I imagine.' Bukit Pelandok rises abruptly to the south, a series of undulating ridges. A bastion in otherwise flattish terrain. The steepness of some of its faces is daunting: how could one even walk up there, let alone with full pack and weapons, or lugging mortar tubes and artillery pieces? Much of the mountain is being quarried, exposing scars of orange soil.

Just east of Pelandok, we take the Medan turn-off and stop where Pilk's men had dug in on that first day. The site is now a sprawling Chinese cemetery.

Along the palm plantation-lined road we reach Yong Peng and imagine the gridlock of beleaguered bodies and transports beating a path for the Causeway. Virtually all Allies had to pass through Yong Peng on the harrowing road south. Then there's an unassuming concrete bridge over the Sungei Bekok River: the replacement of its predecessor blown up by 2/30 Battalion sappers on 23 January 1942 as Westforce made good its retreat. At the Yong Peng–Muar road intersection, a large mosque presides in its blue, white, and gold modern Moorish glory.

Seremban, the capital of Negri Sembilan, is a bustling provincial capital now, full of medium-rises and shopping malls which ring the early-twentieth century shophouses downtown. But the station still presents as a colonial throw-back with its brick clock tower. We almost expect Pilk to greet us at the platform.

The Rubber Planters Association put me onto W.T. 'Pete' Perera, an ex-manager of Ainsdale — Pilk's main estate — who spent 28 years planting from the mid-'60s onwards. The British gave Pete's father an MBE in 1953.'He was a school principal and did a lot of social work, but was some kind of spy for the British during the war,' Pete says. I probe him further on how Malaysians view that period. 'All completely forgotten now ... what has taken place was cruel but nobody talks about it because of lapse of time.' That it was a racially divisive time is one reason Malaysians prefer to gloss over it, almost consider it taboo. Youngsters especially don't understand the need to dwell on such depressing 'old school' times, given there is so much positivism in the new Malaysia.

Perera points us toward Sime Derby's New Labu Estate, where we meet Corporal Nagan, with a row of medals earned from 25 years' service with the estate police, resplendent in his smart blue uniform with bright yellow flip-flops. We ask about Ainsdale Estate, now under their ownership. Nagan was born here, and has lived only on this estate where his father was a cook. When his father retired, he took his dad's job. 'Ah, Pilkington, yes, we had a Pilkington,' he says, rattling off many of the English managers, the last being as late as 1979 despite large-scale 'Malaysianisation' since 1957.

'The English good,' he says. 'Thumbs up, look after the hard workers.' He invites us into his home, a modest dwelling near the entrance gate. He switches on the television. A game of football is in progress. The 53-year-old father of four supports Liverpool and England, and relaxes with a game of badminton followed by some toddy, beer, or a bottle of Chivas now and again if some special friends come over. I probe him on his thoughts toward the Japanese: 'My father tells stories ...' he motions a beheading movement. 'Now, it's different. We are independent. No need to worry, we are the boss. Before, see Japanese we are scared. Now, they come to do business.' Which they do in large amounts: Japan is Malaysia's second biggest source of imports (behind Singapore) and fourth biggest market for exports. Malaysia's total trade with Japan stands at approximately RM100 billion annually, and Japan's annual investment in Malaysia was over the RM1 billion mark even before the 2005 Japan–Malaysia Economic Partnership Agreement was introduced.

Over a ridge at Ainsdale, among the newly planted palm oil, sits a pile of rubble down yonder, only a concrete water tower (which like Pilk we later spot from the train) standing proud – the colonial bungalow home where Pilk and Phyllis enjoyed daily curry lunches prepared by his 'fat and cheerful' houseboy Ah Lim, no longer exists. Pulled down just two weeks prior to our visit. 'I don't know what it means that the bungalow's just been pulled down, but it means something,' says Paul, a little maudlin.

In front of that, in a dip, is the busy Lebuhraya Utara-Seletan (north –south expressway) which runs the entire 1000 kilometre length of the country. 'Well, my dad wouldn't have been too impressed with them wanting to put a highway through in front of the bungalow,' says Paul in a slightly indignant English manner. To the south, beautiful mountains emerge from the morning mist. 'I recognise those mountains from watercolours my mother used to paint.' They are touchstones to his past. As is the train line, visible on the other side of the bungalow. The same line Pilkington used to walk along on estate inspections with his dogs; the very same line that transported him north to the Death Railway and south again to Singapore.

Corporal Nagan, the security guard at the former Ainsdale Estate, Seremban. His father was a cook here, and he remembered the Pilkington family. He and his wife treated us to a huge local home-cooked feast.

View from the former bungalow's position. Paul recognised it from watercolours his mum used to paint.

Institutionally, Malaysia has done nothing to propagate the history of World War II. There is not a single public war museum. Even the iconic Pudu Gaol, built in 1895 by the British, which housed POWs from the Malayan campaign before the fall of Singapore, has been sold for redevelopment. It is now the glitzy Bukit Bintang City Centre, and thankfully the Victorian-era gateway of Pudu has been retained as a nod to its past.

Aside from a few pillboxes along the Mersing Road, only Batu Muang Fort on Penang stands as an intact reminder that a war ever took place. Billed as the largest war museum in South-East Asia this 20-acre British fort was created for £40 million in 1930, only for Penang to be evacuated, and it was lost to the rapacious jungle until 1995 when Tuan Johari excavated it, literally with his own hands (and his own hands in his wallet to the tune of MR16, 000, 000 for the land lease and improvements). Why? 'Because my family had a hard time in the war, it is my duty to bring all this back. So welcome back to Malaya,' says the upbeat clean-cut Malay, as we pass a British flag. 'We admit our colonial master give us everything.' He has, like all his countrymen, served two years of national service. 'It is good for discipline.'

An update on our journey: True to Jo Beer's words, we are feeling rather sorry for ourselves by this point, headaches and an out-of- sorts feeling predominant. Sitting on a train from Singapore to Kanchanaburi for 54 hours doesn't help either ...

Ban Pong, the pre-war Thai railhead, is surrounded by temples, gilt pagodas glinting in the early morning rays. A massive temple complex dominates the skyline. I wonder if that's the one that doubled as a brothel? Otherwise, the sugar-milling centre of Ban Pong has a few prosperous residential pockets but clearly is short on attractions. Half a dozen grey cattle carts sit around the southern end of the station, exactly of the type seen in POW photos. A quick inspection reveals some small numbers on one of them. 11-10-43. A WW2 date? I like to think so. The steel walls are too hot to touch: we could easily fry an egg on its inside panels.

Hard yards up the Railway line from Ban Pong.
This cattle cart at the siding had a date 11-10-43.

View of Kanchanuburi city from the Bridge. 'It was so quiet and silent, you think, "Did that all really happen?" Like a bleedin' dream,' said Dick Lee.

Below: The Wang Po Viaducts were the target of unsuccessful Allied bombing raids. This is the most impressive section of the railway still operating.

We disembark at the Nakhom Pathom junction where the Phra Pathom Chedi, the oldest Buddhist temple in Thailand (dating back to the fourth century) supposedly houses relics of Lord Buddha. Its 127 metre stupa – the tallest in the world – holds sway over the town and was a talking point of all POWs stationed out this way.

A couple of rattling hours later we finally arrive at Kanchanaburi, where the first things that strike me are the logos emblazoned on the station's benches (JTB, a major Japanese tour company). Ironic, for here was Kanburi Camp, one of the largest war-time hospital camps. Kanburi camp is now a nondescript grassy area, like an unused football field, with a library at one end along a busy road that leads parallel to the line through the guts of the laid-back, low-rise provincial capital, home to around 175, 000. Sugarcane and tapioca are the main industries these days. At the immaculately maintained cemetery over the road, 7000 English, Dutch, Australians rest in peaceful rowed sections. Average age, 26.

Little known is that all these graves were dug by Japanese soldiers, post-war. Told to dig them 'as soon as possible', they worked overnight to do so. The story goes that three Japanese suffering from malaria took a break from their digging duties, only to cop a flogging from their senior officer. The Japanese dished it out to their own with equal ruthlessness.

Many of the Royal Norfolks lie together in one area, some as young as 21. Others' names are on the wall, their ashes interred together in Plot IX. We almost literally stumble upon Private Charles Henry Carlton's gravestone, Pilk's batman laid to rest among the Australian contingent probably because he was buried in a mass grave with mainly Aussies. I read Pilk's tribute to his faithful offsider and am overwhelmed by tears. It is a most poignant moment – another feeling of the narratives of past and present colliding.

The Thai–Burma Railway Centre is a private concern adjacent the cemetery, founded and funded by Australian Rod Beattie, who was formerly manager of the Hellfire Pass Memorial Museum and currently oversees the cemetery for the Commonwealth War Graves Commission. 'I've done more than any POW in clearing the Railway.' He and his Vietnamese wife, Thuy, spent four and a half years surveying and reclaiming the tracks from the jungle. 'I'd go out at 6 am, back at 9 pm, with a brush-hook *parang*. I'd eat sardines, toast, eggs. People were wondering if I had AIDS because I lost so much weight.'

During that time he uncovered 60 former POW camp sites. 'When you find, say, an old Japanese bridge, it is very satisfying,' he says of what kept him going. His most poignant find? 'A Chinese goddess of compassion figurine. Someone had obviously bought it in Singapore, looking to send it home, but took it up to the Railway and died of cholera.'

Beattie's manager Terry is also Australian, a living example of how the Railway affected — and still affects — lives. His father Jack Manttan was just 18 when the HMAS Perth was sunk in the Sunda Strait. He never recovered from the privations of the Railway and died of malignant hypertension at the age of 25, the father of three, Terry being just six months old. I get the sense that many grown 60-year-olds (offspring of POWs like Terry) endured childhoods and perhaps adult lives disrupted by the nagging question of why they were robbed of a father. Where is this Thailand place anyway? And how, *why*, did the Railway kill my dad?

One Aussie who has made several trips to this centre and the upper reaches of the Railway is Doug Ogden. His father, Private Jake Ogden of 2/29 Battalion, died at Kami Song Kurai near the Burmese border in 1943, and is buried in Thanbyuzayat. Despite tremendous corporate success, Doug always felt bereft that there were no family photos with all members present. 'There was unresolved grief for me.' For many years his mother would not even walk on the same side of the street as an Asian, affecting Doug by osmosis. Finally in 1996 he stared his demons down and made the pilgrimage to Singapore, Malaysia and Thailand to 'find' his father. (He also joined our party at the Parit Sulong commemoration in 2007.)

'There are no winners in war,' he says. 'Those who served were scarred, those left behind were scarred. For what?' Doug, who I met while doing the Sandakan Death march trek, has been consistently active with the 2/29 Battalion association. 'I've had a lot of anger and resentment about the war having grown up without a father.' Tears well in his eyes.

'And they're still sending our young people off to fight silly wars,' says his teary wife, Liz.

He has made separate pilgrimages with his daughter, his brother, and Beattie. 'I've had some wonderful experiences with Rod, bashing through the jungle finding where some of Dad's camps were. Sometimes I say "No more!" but someone asks, or I go to the Shrine, and I'm off again.'

I ask Beattie — who comes from a family where many served their country — his take on the Japanese military mindset that produced so much atrocity. 'A civilian can *never* understand a military mind,' Beattie says emphatically. 'They were acting under orders, simple as that. They don't ask why.' One thousand out of 15, 000 Japanese and Koreans assigned to the Railway died here, he tells me. Nearly seven percent. 'That's a hell of a death toll for a non-combat situation, but the task was to build the Railway. But where does the blame lie for all this?' asks Beattie. 'How far up do you go?'

Peter is a canon from country New South Wales. His father was on the Railway. 'I can't forgive them,' says this otherwise good-humoured man of the cloth. 'It still makes me want to smack them, so can you imagine how the guys who went through it felt? It's bloody hard to be civil to a Jap after all that.'

Kanchanaburi is a buzzing tourist town, popular as a fun getaway with the Thais, who come for the river cruises, karaoke, Tiger Temple ... and, of course, the Bridge on the River Kwai. But how much do they actually know about World War II? I canvas the history skills of a few locals. Malit, about mid-20s, has a wonderfully sunny disposition. Her history knowledge, however, is a bit cloudy: 'We have too much history in Thailand. We read about it in school, then forget about it. Maybe at university they learn more.' Her implication seems to be that Hellfire Pass is of interest to Westerners only. The assistant manager at the Hellfire Pass Memorial Museum, Ayr, is one of the best advertisements for 'the land of smiles' around. Originally from the south, she was raised in Bangkok. I quiz her on what she knew about this episode before taking the job: 'Growing up, no, we know nothing about this history ... we rather go shopping'. She laughs. 'Even when I accept this job — I was in Australia at that time — I thought I was going to be in Kanchanaburi city but my mother called me to say, "You crazy,

Hellfire Pass is in the jungle, 80 kilometres from Kanchanaburi".'

Hellfire Pass is known in Thai as Chong Khao Khart (Broken Mountain Pass). 'C.K. Pass ... have army before,' answers pony-tailed Orn more than a little sketchily. But her friend Au is from Kanchanaburi. 'People from other parts don't study about this, but if it's our local history in this area, yes, we study about it in school.' She says she's been there, and says it gives her a creepy feeling. 'Like *pii*,' she says, rubbing imaginary gooseflesh. Ghosts.

She is very perceptive. Rod Beattie tells me of working overnight making preparations for the ceremony to scatter Sir Edward 'Weary' Dunlop's ashes here in 1994. 'I looked up and saw these ghosts. About four or five POWs along the top of the cutting, just having a casual look down at me to see what I was doing.'

The highlight of our trip to me is going up the original railway way line from Kanchanaburi, over the Kwai Bridge, along the Wang Po viaducts, then hopping off at Nam Tok (formerly Tarsao), which is now the railhead. Rod Beattie tells us that Pilk's account of his and his band of infirm POWs being transported *north* to work camps by rail is the first he's come across.

The Bridge on the River Kwai is not as heroic-looking as it was made out to be in the movie (filmed in Sri Lanka, incidentally, as Thailand was deemed not to be jungley enough). Many bars screen the movie nightly, and the most resonant line in it is when the officer — Colonel Philip Toosey in real life, played by Alex Guinness — says; 'I hope in years to come, when the war is over, people remember who built it and how they built it.' Amen again. Ironically, not by watching that movie. We can lay the blame for that — and understand Pilk's dismissal of it — on French author Pierre Boulle who was a POW but not in that area. His story was based on hearsay, and many veterans called for the resulting movie to be labelled as fiction. Toosey's grand-daughter, author Julie Summers, tells me Toosey did not even recognise his character in the film, which Dick Lee reviews as 'a load of shite'. (I was pleased to meet Julie in 2006 when she was promoting *The Colonel of Tamarkan* in Australia, and I promptly organised a evening for her to meet with Lynette Silver, and I sat revelling in their collective knowledge and passion. I was also thrilled to attend Julie's launch of *Stranger in the House* in London, 2009, and spent some time in her lovely home in Oxford.)

'Knowing Toosey pretty well, I was disgusted by it,' says former Bombardier Allan McNevin of 2/10 and A Force about the *Kwai* movie. 'It was good entertainment, though,' he concedes.

However wide of the mark, the movie has done more than anything else to propagate awareness of the Death Railway.

On the bridge I start whistling the 'Colonel Bogey March' unconsciously, the infectious theme song of *The Bridge on the River Kwai*. From behind me, I hear another whistler join in. I turn to see a middle-aged Japanese man and his wife. 'Good movie, huh?' I say, and he gives me a broad gold-toothed grin. It occurs to me how much anti-Japanese sentiment builds up when you get immersed in this story.

There is a special tourist train that plies this route, but we eschew it. Instead, we talk our way into the 'staff and monks' carriage for the two hour sector to Nam Tok. Spartan wooden trestle seats and rain slashing in via the open sides. Perfect! The Wang Po viaducts beggar belief: the sheer volume of rock removed to create the ledge along the cliff edge, and the skill in creating the timber trusses. While these have been restored and upgraded over the years, it's largely unchanged from 1943, and the train slows to a crawl and wends its way ever so gingerly — amid creaks and groans of protesting logs — across the viaducts, affording a splendid view of the serpentine Kwai Noi river. At Wang Po station, two large bomb craters straddling the track are clearly visible. Had these struck the viaducts it might well have been game over for the Japs' use of the railway.

Bamboo thickets and teak trees compete in a death match with vines for sun and survival. The tang of lemongrass as the train whips its way through overhanging vegetation is refreshing. Electric-blue kingfishers eyeball the train. Misty peaks soar seemingly thousands of feet over the river to the west.

The terrain in this area is horrendously rugged, and it defies comprehension that anyone could've thought it feasible to run a train line through here. The road, river and train track form a holy trinity of sorts, criss-crossing each other like a plaited vine. The original Japanese marching route is now the wide and wonderful Highway 323 up to the Burmese border.

Nam Tok, formerly Tarsao, is the end of the line. At the end of the war, the Allies decided it would be too costly and difficult to dismantle the railway. They ripped up a four kilometre section near Three Pagodas pass around the Burmese border in order to stop it being used to feed Karen separatists. Instead, they briefed Lieutenant Colonel Warmerhoven – a Dutch officer and railway enthusiast who had taken it upon himself to take over the running of the line the day after the Japanese surrender – to value its assets. The line was sold to Thailand for £1.25 million. Some of this was distributed to POWs in 1952.

The section from Nong Pladuk to Nam Tok was closed from 1947 to 1956. With operation of the whole line economically unfeasible, the section from Namtok to the Burmese border was dismantled. Forty kilometres of that route is now submerged beneath the waters of Khao Laem Dam. The remainder was gradually upgraded and 130 kilometres from Nong Pladuk to Namtok remain in use.

Nam Tok is a busy little town abutted by the Burma Ranges, its main tourist appeal being the Sai Yok Noi waterfall, located along the former track. Here, a Japanese locomotive sits in the cutting. On the edge of town, a market of canvas stalls peddles fish, fried crabs laced with chilli, chicken and pork satays, great woks sizzling away and cauldrons bubbling. This is sensory torture to Paul and me, on our Death Railway rations, and by this time we are generally feeling woozy and light-headed. Weight loss is already evident, taking my belt in a notch or two, and we find it difficult to consume all the rice each day, often leaving half aside. Our stomachs seem to have shrunk.

A pond and grand entrance towers lead to the Pung Waan Resort, which now occupies the site of the former Tarsao Hospital Camp, the HQ and hospital for D Force. West of the lily pond near the entrance was where the Tarsao Work Camp was located. It is now beautiful bushland. The spacious resort – still owned by the same family, made good by successful trading business in Bangkok – has springbok and other deer aplenty in its lush orchard-laden surrounds. Guests take their time over a coffee, swim, or sign up for a soothing massage. It is hard to fathom that nearly a thousand POWs were buried here. All were subsequently exhumed and laid to rest in Kanchanaburi.

Pilk made the point that, say, in a group of four POW friends, if one died the other three would soon follow. All told it was easy to die, harder to live. I'd asked Dick Lee – who'd had 'a bleedin' rough time here at Tarsao' – how much of survival was mental fortitude:

'You did get people who did chuck the towel in, the odd few who give up too early. I had amoebic dysentery and I thought I was a goner then – because they were dying like bleedin' flies. You couldn't get the bleedin' injections, you couldn't get treatment for it, and all of a sudden I'm burying people, dozens at a bleedin' time during the week. Funeral going out with a Japanese guard to the area where they dug up as a graveyard sort of thing. Then all of a sudden you go down with it. You're going thirty-odd times during the night. Unbelievable. 'Cause it's only water and blood coming from ya. You're just wasting away, you lose stones in a couple of weeks. Normally I was 10.5. I went down to six stone.'

Was there one thought in his mind that kept him going? 'Yeah, in your mind you say, "These bastards aren't going to get me".' Many veterans tell me the same thing: 'I'm not going to die in this god- forsaken hellhole, no way'.

Unfortunately we don't permit ourselves to stay at Pung Waan, but check in at a less salubrious place along the main road. Having covered around 20 kilometres on foot in the area, and with over a week of accumulated body odour, Paul insists I shower. I accuse him of being on an Abercrombie and Kent luxury tour. He counters by saying I'm taking this realism thing too far. Putting on the same damp, smelly clothes after a shower is both a let-down and anti-social ... I want to feel sharp, crisp and fresh again. We would kill for a Singha beer to accompany our view of the sunset that evening. Despite no alcohol, we both feel dizzily drunk, very slow in forming thoughts, and finding the words is difficult. 'It's bad enough in English let alone other languages,' says Paul, sprouting serious whiskers by now.

Later there is a knock on the door: 'For you,' says the kitchen hand, kindly proffering sliced pomelo. We glance at each other, then unashamedly dive in.

The pelting rain of the lingering wet season compounds the

misery of our 5 am tenkos (roll calls). Hard to motivate yourself when, as in Kanchanaburi, other guests at the $3-a-night hostel were just dragging themselves back from a long night on the town. All these things begin to weigh heavily on us. I am irritable, tearing at my fingernails; my hair-trigger fuse fires up and I snap when Paul announces he's lost the diary. 'Only 67 hours of this mission to go,' sighs Paul as we prepare to walk to Hellfire Pass.

We trek Highway 323 in search of the former Tampii camp site, armed with mud-maps (literally) hand-drawn by Rod Beattie. A school and temple signify its location – Tampii camp straddled the track – confirmed by the flat field and presence of a small stream. At the back of the school, no sign of the Railway; just a peaceful pond. The line in fact cut straight through – north–south – the current school building which sits on the site of the original Tampii station. A little trail leads off north. A young couple in a 'Save the Wildlife' van assure us this is the route to Hellfire Pass. 'Maybe 10 kilometre.'

Roadside stalls shift melons, pineapples, bags of chillies, bottles of bamboo shoots, ginger. Further along, stopping for some water, the aromatic steam from a wok of pork, garlic and chilli blows into my face.

The conical mound of Malay Hamlet camp is immediately north of the signed turnoff to Hellfire Pass. Instead of 'the nauseating smell of burning bodies' Pilk experienced, the humid air is instead perfumed by a nursery with dazzling blooms near the gurgling stream. The area is swathed in tall yellow grass and thick bamboo stands backed by jagged cliffs. We walk in a fair way but come across nothing of interest (although Beattie told us a pile of burnt remains of cholera victims' clothing was recently found here).

I try to strip all the vegetation out of the image and replace it with the huts and graves on the sloping hill ... just as Pilk knew it. Poor, poor bastards. And to think the epidemic could have been prevented as the Japanese had stockpiles of cholera vaccines; it's just they never made them available.

'It's a fair way from the Railway itself,' comments Paul. Indeed, POWs often faced a march of several kilometres just to get to their dispiriting workplace. Priority was given to ease of supply, flat ground and water source.

Hellfire Pass is situated within Royal Thai Army land. Immediately we are faced with an open field, the area where the notorious Kanyu 3 camps were situated. A sign next to a restaurant of sorts proclaims: 'HELLFIRE STEAK. Drink to your health.' I find this tacky. Completely inappropriate. (This was subsequently changed in 2006 to Hero's Garden.)

At Hellfire Pass Memorial Museum, manager Bill Slape – in army-style pants and police boots – explains they're getting more and more Thai youth along here now that they offer an MP3 tour. 'Part of the attraction is that it's called Hellfire Pass ... hell and death are taboo to them normally but, because it's called Chong Khao Khart in Thai, it's not taboo any more.'

Although Japanese represent only a small percentage of visitors, many Japanese visitors apparently walk the trail. 'They make the effort to learn,' says Bill, who suspects many former Japanese guards also come back here: 'They're older blokes, they don't say anything, but you suspect it. Reinichi Sugano, the senior railway engineer with 9 Railway Regiment who drove the whole project, has come back a few times. 'He's absolutely not remorseful. He brought his wife once – she was a frail old thing and she looked around and, as she went to leave, she fainted. To her he was obviously all hearts and flowers, and hadn't prepared her for what he really did.'

(Bill sadly passed away from a heart attack in 2012, soon after he and local wife Aor, the very capable manager at the museum, had built a nice home in Kanchanaburi. I had myself moved to live in Thailand from 2009-2020 and an annual highlight for me was to motorcycle to the Anzac Day ceremony, when a wholly excellent service was put on in the Hellfire Pass cutting each year.)

From the Museum's viewing platform, over a huge bowl of serene floating orchid petals, the Burma Ranges are folded, like origami, three or four, or more, deep. Shrouded in mist and mystery. The stands of teak and bamboo are achingly beautiful.

'I walked into the jungle where a bit of the old railway was still lying,' Dick Lee tells me. 'And it was so quiet and silent, you think, "Did that all really happen?" Like a bleedin' dream. I couldn't realise that I'd been there.'

In the cutting is a plaque honouring Aussie POW surgeon

Sir Edward 'Weary' Dunlop whose ashes are scattered here. A controversial new black granite cairn sits on a circular plinth. It recently replaced the temporary cairn and wooden cross traditionally used for memorial services. Sleepers have been repositioned, and ballast laid and prettied up. A skip sits atop the cutting on a section of rail. 'When they rearrange things it becomes like Disneyland,' says Paul, disillusioned. 'It means nothing to me then. And when they pretty it up and add things you're not sure what to believe any more.'

The 600-metre stretch of red-rocked Hellfire Pass is truly humbling. But, pushing north, the scope and scale of the Railway imposes itself. The embankment built by hand with countless cubic metres of soil and rock. Hintok, where Ray Parkin observed the two malarial POWs assisting their cholera-ravaged mate. Ravines and inclines, hundreds of feet deep and crossed by bridges, famously the Pack of Cards Bridge and Three-Tier Bridge. Sadly, nothing remains of these massive bits of workmanship.

Modern engineers marvel at the Death Railway as an engineering feat. The rugged, inhospitable terrain, the lack of tools, the superhuman physical effort, and, yes, the know-how of the Japanese engineers. Some of the best minds behind the Thai–Burma Railway – including engineer Yoshihiko Futamatsu who directed construction – went on to design Japan's record-breaking Shinkansen 'Bullet Train' system, characterised by its use of tunnels and viaducts to go through or over obstacles rather than around them. Sounds familiar ...

In the middle of the awe-inspiring Hellfire Pass cutting. Time here was always a highlight for those in my tour groups as the enormity and atrocity of the Death Railway really sank in. It is also the site of the annual Anzac Day dawn ceremony, where the two-minutes silence is filled with birdsong.

Paul pays his respects to his father's loyal batman, Carlton, in Kanchanaburi War Cemetery, 2005.

I completely lost it when reading out Pilk's tribute to Carlton there.

So did the POWs derive any satisfaction (no matter how guilty) from their role in this engineering achievement?

'Well, my father didn't have any pictures of it on the mantelpiece, if that's what you mean,' says Paul.

'When you look back at it all, they were no mugs,' says Dick Lee. 'The Japanese knew exactly what to do, they were good engineers, and they had the right people out there to do the job.'

'Any pride or satisfaction? No, maybe for others but not for me,' says John Miller who was at Konkoita when it joined. 'No,' says Allan McNevin, also at Konkoita. 'I can't say there was any great feeling. Just another's day's job then back to the grindstone. There was still plenty to do laying ballast and packing behind it. The unballasted rails looked like a snake in the heat of the day, and had to be straightened with a crowbar. A feeling of relief that the pressure would drop, the panic would be over. But no great feeling of We've Done It. No bottles of champagne came out, put it that way.'

'No!' confirms Bill Haskell of the Australian 2/3 Machine Gun Battalion, rather angrily. 'In fact, a lot of *dissatisfaction*. There were too many lives lost, too much brutality. It was a good job it was over. I'm glad we didn't have to do another monsoon season — that cholera ...'

Our mission, too, is thankfully over. Back in Bangkok, freshly showered and sporting a fresh change of clothes. Paul and I find our dreams of a massive meal of scrambled eggs, toast and coffee has evaporated. Instead we celebrate with a tuna sandwich and hot dog from 7-11. I find I have lost twelve kilograms from this exercise, and feel thoroughly detoxed.

We find a cosy bar where we drink our fill of crisply chilled Heinekens and dance like fools. It just feels like the right thing to do at the time. There's a Japanese gent in the bar, with silvering hair and specs. Ito smiles and asks 'Why are you here?' Oh gee, do I tell him the whole Battle of Malaya/Death Railway thing, or do I just dodge it. I dodge it. 'Just for the beer!' I say, and this finds great favour with him. He asks where I am from. Sydney, I tell him.

'Sorry for bombing Sydney Harbour with submarines,' he says, apropos of nothing. I nearly choke on my mouthful of beer. 'But that's not me, that's my father.' It turns out 58-year-old Ito, who now works in 'press images' — (hmm, a likely cover I'm thinking) — was in the Japanese Navy for 25 years. 'We drink to peace,' he says, and we clink glasses. He has two children in their twenties. 'My job is to tell them that war is bad,' he philosophises.

As he's introduced the topic, I weigh in. Does he know about the whole Death Railway saga? It turns out he knew the story from school and has visited there. So what does he make of the Japanese prime minister wanting to change the Japanese constitution away from a pacifist one. 'He half crazy,' he says, tapping his temple, 'like George W. Bush.' We laugh, clink glasses, and drink to peace. Repeatedly.

This mission has taught me much about the South-East Asia of today, the Malayan campaign, the POWs' lives after the war, the Pilkington family, but — it seems — most about myself.

For all my fascination with military history, and especially military 'toys' and technology, I finally resolve — after years of vacillating ambiguity — that I am now firmly, undeniably, resolutely anti-war. I can no longer view Kranji or Kanchanaburi — or any other war cemetery for that matter — in the abstract. Every gravestone, should the victim be lucky enough to have got one, is not just a casualty or statistic: it is someone's son or daughter, someone's flesh and blood, a bundle of hopes and dreams cruelly and prematurely dashed. Looking at my own wonderful children, I can't begin to imagine how devastating that loss would be.

At the time of writing my son is submitting his application for the Australian Defence Force Academy. He wants to be an armoured corps officer. Tanks, in other words. That's OK, because you can't use tanks in Malaya, right? *Jungles are much too impenetrable, old chap.*

Major Kevin Fagan, one of 43 Australian doctors on the Railway, attended the same school as my son and I, so Pilk's reverence of him has a poignant resonance. A boarding wing bearing Fagan's name was recently opened at Riverview College, which I had pleasure showing to Paul. This quote from Fagan on being a prisoner of the Japanese is inscribed on a plaque in Hellfire Pass: 'It gave me a great understanding of men. And a great appreciation of the ordinary things of life ... and the value of human relations. You know, when it comes to the end, the only thing that really matters are the people whom you love, and who love you.' (In 2009 I would do a book launch talk in the Kevin Fagan Wing. Very poignant.)

As the saying goes: if you haven't supped from the same cup, you can't possibly know the taste. After what amounted to a dreadful detox, I've drunk only decaffeinated coffee and gave up beer (although red wine has proved a wonderful substitute). But it is all I could do to empathise even in the slightest, say a millionth, with the plight of the POWs. I have since become somewhat obsessed with the story, indeed the myriad stories, of the Malayan campaign, which has seen me leading tours of veterans' relatives and other history buffs through Singapore, the battlefields of Malaya, and up the Death Railway, where each time I live in hope that the jungle will yield up more answers. Among other things, I've learnt that going into the war Pilkington might have been the same weight as I, but he was clearly a far bigger man in every respect than I'll ever be.

Captain Pilkington, sir, I salute you and all other POWs held by the Japanese. Because now -- after all this -- I've got even *less* idea of how you survived The Missing Years.

-- *The End* --

If this story moved you, please be kind enough to share your thoughts in a review on Amazon or on Goodreads so others can learn from this experience, too. Thanks!

Acknowledgments (updated 2021)

It has taken me approximately the same length of time to write and research this book as the beleaguered POWs whose story is told here spent as guests of the Japanese. I know in which way I would prefer to have spent my time: some wonderful friendships and alliances have been forged, for which I am truly grateful.

Firstly, to the Pilkington family for allowing me to intrude so deeply and so regularly into their private life. This bond has only deepened in the decade or so since.

To David and Scilla Rosenberg, the original publishers of this book. Thanks and congratulations on your well-earned OAM award. Enjoy your well-earned retirement. Thanks to Carl Harrison-Ford whose deft editing cut away about as much rock as was removed from Hellfire Pass. Military historian par excellence and author Lynette Ramsay Silver, AM, punished me unto near death on the Sandakan Death March so I sought revenge by getting her to fact check this manuscript.

Before any of that happened, a mountain of gratitude to the indefatigable Rod Beattie and Terry Manttan at Thai Burma Railway Centre who every day help veterans and relatives make sense of the Railway and still make time for interlopers like me. And to Andrew Snow who, with sister Tina, joined one of my military history tours subsequent to the original publishing of this book. Upon return to Australia, he sold up and left Sydney to become a volunteer at the TBRC. But not before he sent me a note saying 'Your tour has been life-changing for me.' And *he* has now become *my* go-to expert on The Railway and is exceedingly generous with his time and knowledge.

To Anne Bennie at the Australian War Memorial for seeing the historical value in this project and facilitating photographic services accordingly, as well as Jeremy Richter and staff at the AWM reading room. Ditto Ong Wei Meng at the Singapore Archives. Thanks to Monica Haynes at Melbourne University Press for negotiating permission from John Parkin for use of his father's evocative cover image, which possibly says more in a few strokes than 100, 000 words of my text.

It was a wonderful moment of serendipity when I found an image of the Sime Road Camp I wanted to use. The Alex Turnbull Library in New Zealand referred me to the image copyright owner, one David Tompkins. That name sounded familiar to me. It turned out indeed I had crewed on his yacht one summer in Auckland 1986 and we reacquainted ourselves. He kindly sent me a copy of his aunt Gladys' book *Three Wasted Years* documenting her time as a nursing sister and POW in Malaya and Singapore, containing more of her glorious watercolours.

The following roll of honour is just a small token of appreciation to people who helped me in ways large and small but without whom I might still be looking for answers in the wrong places: Andrew Kefford; Australian-Thai Chamber of Commerce: Peter Roennfeldt and Martin Kyle; Alexandra Hospital: Tan Din Xiang and Ang Kee Seng; Ian Boyle; Changi Chapel and Museum: Jaya and guides; Cyril Gilbert; Devika Misra; Diane Johnston; Dick Lee (RIP, sir); Hellfire Pass Memorial Museum: Bill Slape (RIP) and Ayr Steer; The Imaginative Traveller: Warren Smith and teams in Sydney and Bangkok; John Kennedy; Dr Jia Qingyan; Media: including Matthew Bellotti (*Malaysia Expatriate Lifestyle*), Duncan Gardham (*The Telegraph*, UK), Denis Gray (Associated Press), Susan Kurosawa (*The Weekend Australian*), Graham Mabury and Kim Hutchinson (6PR), Rod Moran (*The West Australian*), Katherine Purcell (2XX), Major John L. Raybould (*Britannia and Castle*), Di Sattler (*The Sunday Times* magazine), *South China Morning Post* magazine, *Sydney City Weekly*; Jackie Neo; Julie Summers; Kuringai Library, Gordon; Nathan Hunter; NRMA Travel: Michael Leary, Cathy Garcia and Rose Toohey, and all those who've joined me on tours through South-East Asia; Doug and Elizabeth Ogden, and the 2/29th Battalion Association; Peter Buckman, The Buckman Agency; Prisons Department Malaysia: TPP Mussa and ASP Letchumanan; Revitalise Pty Ltd: Jo Beer; Rod Eime; Royal Norfolks Museum: Kate Thaxton; Rubber Planters Association: Jane Lopez and W.T. Perera; Sai Yok Yai National Park rangers. Shannen Ho; Singapore Armed Forces: public relations team and assigned guides at Roberts and Selarang Barracks;

Singapore Tourism Board: Carrie Kwik, Sandra Leong, Michelle Alice Thoo and Amity Roche; Stella Clare Wee; Tourism Authority of Thailand: Pongsak Kanittanon; Tourism Malaysia: Anne Aziz-Parker and Lyn Tan.

War Museum, Penang: Tuan Johari Shafie; Woon Lee; and Zen Tan.

Bibliography

Allbury, A.G., *Bamboo and Bushido*, Robert Hale, 1955.

Allen, Charles, *Tales From the South China Sea*, Abacus, 1994.

Aspinall, George, and Tim Bowden, *Changi Photographer*, ABC Enterprises

Barber, Noel, *Sinister Twilight*, Cassell Military Paperbacks, 2002.

Bayly, Christopher, and Tim Harper, *Forgotten Armies*, Allen Lane, 2004.

Beattie, Rod, *The Death Railway: A Brief History*, Image Makers, 2006.

Boyle, James, *Railroad to Burma*, Allen & Unwin, 1990.

Braddon, Russell, *The Naked Island*, Pan Books, 1955.

Brazil, David, *Insider's Singapore*, Times Editions, 2001.

Cameron, Nick, *The Complete SAS Guide to Safe Travel*, Piatkus, 2002.

Christie, R.W. (ed.), *A History of the 2/29 Battalion – 8 Australian Division AIF*, High Country, 1983.

Church, Peter, *A Short History of South-East Asia*, Asean Focus Group, 1999.

Daws, Gavan, *Prisoners of the Japanese*, Scribe, 2008.

Elphic, Peter, and Michael Smith, *Odd Man Out: The Story of the Singapore Traitor*, Hodder & Stoughton, 1993.

English, Jeffrey, *One for Every Sleeper: The Japanese Death Railway Through Thailand*, Robert Hale, 1989.

Ferguson, Niall, *Empire*, Allen Lane, 2003.

Frei, Henry, *Guns of February: Ordinary Japanese Soldiers' Views of the Malayan Campaign and the Fall of Singapore, 1941–*

42, Singapore University Press, 2004.

Gardenwise, Singapore Botanic Gardens newsletter, Vol 25, Singapore National Parks.

Hall, E.R. ('Bon'), *The Burma-Thailand Railway of Death*, Graphic Books, 1981.

Holmes, Linda Goetz, *4000 Bowls of Rice*, Allen & Unwin, 1994.

Kirby, S. Woodburn, *History of the Second World War: The War Against Japan*, HMSO, 1957.

Lack, John, *No Lost Battalion: An Oral History of the 2/29 Battalion AIF*, Slouch Hat, 2005.

Lee, Cecil, *Sunset of the Raj*, Pentland Press, 1994.

Lee, Donald, *A Yarn or Two*, Hesperian Books, 2000.

Lloyd, Stuart, *Hare of the Dog*, Captions, 2003.

Lomax, Eric, *The Railway Man*, Vintage, 1996.

Nelson, Hank, *Prisoners of War: Australians Under Nippon*, ABC Enterprises, 1985.

Partridge, Jeff, *Alexandra Hospital: From British Military to Civilian Institution 1938–1998*, Alexandra Hospital/Singapore Polytechnic, 1998.

Peek, Ian Denys, *One Fourteenth of an Elephant: A Memoir of Life and Death on the Burma-Thailand Railway*, Doubleday, 2004.

Richards, Rowley, *A Doctor's War*, HarperCollins, 2005.

Searle, Ronald, *To the Kwai and Back: War Drawings, 1939–1945*, Souvenir Press, 1986.

Shinozaki, Mamoru, *Syonan – My Story*, Marshall Cavendish, 2006.

Silver, Lynette Ramsay, *The Bridge at Parit Sulong*, The Watermark Press, 2004.

Stewart, John, *To The River Kwai: Two Journeys, 1943 and 1979*, Bloomsbury, 1988.

Summers, Julie, *The Colonel of Tamarkan: Philip Toosey and the Bridge on the River Kwai*, Simon & Schuster, 2005.

Sutherland, Jon, and Diane Canwell, *World War II: A Source Book*, Flame Tree Publishing, 2003.

Tett, David, *A Postal History of the Prisoners of War and Civilian Internees in East Asia During World War Two*, Volume 1: *Singapore and Malaya 1942–1945, The Changi Connection*, BFA Publishing, 2002.

Tsuji, Masanobu, *Japan's Greatest Victory, Britain's Worst Defeat: Capture and Fall of Singapore, 1942*, Spellmount, 1997.

Turnbull, C.M., *Dateline Singapore*, Singapore Press Holdings, 1995.

20th Century Day by Day, Dorling Kindersley, 2000.

Wall, Don, *The Heroes of F Force*, self-published, 1993.

Wigmore, Lionel, *Australia in the War of 1939–1945: The Japanese Thrust*, Australian War Memorial, 1957.

Wilmot, H.P., *The Second World War in the Far East*, Cassell Military, 2002.

Yap, Siang-yong, *Fortress Singapore: The Battlefield Guide*, Singapore Times Editions, 1998.

Personal Discussions and Interviews

My undying respect and gratitude to these veterans who shared all or part of their story formally or informally with me to further my understanding of certain episodes:

Dr Bill Frankland, RAMC; Cyril Gilbert, OAM, Sergeant, Australian Army Service Corps, 8 Division; Rear Admiral Guy Griffiths (retired), Royal Australian Navy; Bill Haskell, 2/3 Machine Gun Battalion; Richard Lee, HQ, Royal Artillery, 11 Division; Allan McNevin, 2/10 Field Regiment and 'A' Force; John Miller, 2/10 Field Regiment; Roy Miller, Royal Navy; Nick 'Albert' Mitchell, 5th Royal Norfolks, 18 Division; and Stan Sharpley, 2nd Loyal Lancashires.

Manuscripts

Bruton, Peter, 'The Matter of a Massacre, Alexandra Hospital Singapore 14–15 February 1942', 1985.

Kuo, Huei-ying, 'Nationalism Against its People? Chinese Business and Nationalist Activities in Inter-War Singapore, 1919–1941', City University of Hong Kong.

Pilkington, Hugh P., 'Bungalow Days', 1944.

Pilkington, Hugh P., 'Eyewitness Account', Alexandra Military Hospital, Malaya, Robert Henry Mutton.

Pilkington, Hugh P., 'Ornot's Ornithology', 1944.

Pilkington, Hugh P., 'Rubber: Its Conditions, Reconstruction and Control', 1931 report.

Pilkington, Hugh P., 'Silver: Paul's Own Story', 1945.

Pilkington, Hugh P., 'The Missing Years', 1945.

Pilkington, Hugh P., Letters 1940 –45.

Varshney, Ashutosh,'Ethnic Conflict and Civil Society: Malaysia, Sri Lanka, Nigeria and Indonesia', University of Michigan, 2002.

Webliography

Australian War Memorial, www.awm.gov.au.

Australia's War 1939–1945, www.ww2australia.gov.au.

Battlefield Singapore, www.s1942.org.sg.

Britain at War, www.britain-at-war.org.uk.

The Britannia and Castle, www.norfolkbc.fsnet.co.uk.

Cambridge Janus Archive, janus.lib.cam.ac.uk.

Commonwealth War Graves Commission, www.cwgc.org.

Far Eastern Heroes, www.far-eastern-heroes.org.uk.

Far East Prisoners of War Community, www.fepow-community.org.uk.

Hellfire Pass, www.hellfirepass.com.

Imperial War Museum, www.iwm.org.uk.

Malaysian Railways, www.ktmb.com.my.

Malaysian Rugby Union, www.mru.org.my.

Pilkington Glass, www.pilkington.com.

Sheringham, North Norfolk, www.at-sheringham-norfolk.co.uk.

Singapore Archives Online, www.a2o.com.sg.

State Railway of Thailand, www.railway.co.th.

Thai Burma Railway Centre, www.tbrconline.com.

The U-Boat War 1939–1945, www.uboat.net.

World War II Multimedia Database, www.worldwar2database.com.

For more bonus material, please visit
catmatdog.com/themissingyears

About Stuart Lloyd

Stuart Lloyd was born in Rhodesia (now Zimbabwe) and has spent one-third of his life in Africa, one-third in South-East Asia, and one-third in Australia. His great-grandfather, George Frederick Seymour, was one of seven officers to ride in with Lord Kitchener to relieve the siege of Mafeking. Apart from 16 books on military history, colonial history and travel, Stuart has been published in *The Australian, Sydney Morning Herald, National Geographic Traveler, South China Morning Post* and *New Zealand Herald*, among others, and has lead military history tours through Singapore, Malaysia and Thailand.

See more at catmatdog.com/stuart-lloyd-author

Made in the USA
Monee, IL
04 November 2026

67386618R00184